IBM Cloud Pak for Data

An enterprise platform to operationalize data, analytics, and AI

Hemanth Manda

Sriram Srinivasan

Deepak Rangarao

BIRMINGHAM—MUMBAI

IBM Cloud Pak for Data

Copyright © 2021 Packt Publishing

Publishing Product Manager: Ali Abidi
Senior Editor: Roshan Kumar
Content Development Editors: Athikho Sapuni Rishana and Priyanka Soam
Technical Editor: Manikandan Kurup
Copy Editor: Safis Editing
Project Coordinator: Aparna Ravikumar Nair
Proofreader: Safis Editing
Indexer: Pratik Shirodkar
Production Designer: Aparna Bhagat

First published: October 2021

Production reference: 2221021

Published by Packt Publishing Ltd.
Livery Place
35 Livery Street
Birmingham
B3 2PB, UK.

ISBN: 978-1-80056-212-7

www.packt.com

Contributors

About the authors

Hemanth Manda heads product management at IBM and is responsible for the Cloud Pak for Data platform. He has broad experience in the technology and software industry spanning a number of strategy and execution roles over the past 20 years. In his current role, Hemanth leads a team of over 20 product managers responsible for simplifying and modernizing IBM's data and AI portfolio to support cloud-native architectures through the new platform offering that is Cloud Pak for Data. Among other things, he is responsible for rationalizing and streamlining the data and AI portfolio at IBM, a $6 billion-dollar business, and delivering new platform-wide capabilities through Cloud Pak for Data.

Sriram Srinivasan is an IBM Distinguished Engineer leading the architecture and development of Cloud Pak for Data. His interests lie in cloud-native technologies such as Kubernetes and their practical application for both client-managed environments and Software as a Service. Prior to this role, Sriram led the development of IBM Data Science Experience Local and the dashDB Warehouse as a Service for IBM Cloud. Early on in his career at IBM, Sriram led the development of various web and Eclipse tooling platforms, such as IBM Data Server Manager and the SQL Warehousing tool. He started his career at Informix, where he worked on application servers, database tools, e-commerce products, and Red Brick data warehouse.

Deepak Rangarao leads WW Technical Sales at IBM and is responsible for the Cloud Pak for Data platform. He has broad cross-industry experience in the data warehousing and analytics space, building analytic applications at large organizations and technical presales, both with start-ups and large enterprise software vendors. Deepak has co-authored several books on topics such as OLAP analytics, change data capture, data warehousing, and object storage and is a regular speaker at technical conferences. He is a certified technical specialist in Red Hat OpenShift, Apache Spark, Microsoft SQL Server, and web development technologies.

About the reviewers

Sumeet S Kapoor is a technology leader, seasoned data and AI professional, inventor, and public speaker with over 18 years of experience in the IT Industry. He currently works for the IBM India software group as a solutions architect Leader and enables global partners and enterprise customers on the journey of adopting data and AI platforms. He has solved complex real-world problems across industry domains and has also filed a patent in the area of AI data virtualization and governance automation. Prior to IBM, he has worked as a senior technology specialist and development lead in Fortune 500 global product and consulting organizations. Sumeet enjoys running as his hobby and has successfully completed eight marathons and counting.

Campbell Robertson is the worldwide data and AI practice leader for IBM's Customer Success Group. In his role, Campbell is responsible for providing strategy and subject matter expertise to IBM Customer Success Managers, organizations, and IBM business partners. His primary focus is to help clients make informed decisions on how they can successfully align people, processes, and policies with AI- and data-centric technology for improved outcomes and innovation. He has over 25 years of experience of working with public sector organizations worldwide to deploy best-of-breed technology solutions. Campbell has an extensive background in architecture, data and AI technologies, expert labs services, IT sales, marketing, and business development.

Table of Contents

Section 2: Product Capabilities

3

Collect – Making Data Simple and Accessible

4

Organize – Creating a Trusted Analytics Foundation

5

Analyzing: Building, Deploying, and Scaling Models with Trust and Transparency

6

Multi-Cloud Strategy and Cloud Satellite

7

IBM and Partner Extension Services

8

Customer Use Cases

Section 3: Technical Details

9

Technical Overview, Management, and Administration

10

Security and Compliance

11

Storage

12

Multi-Tenancy

Other Books You May Enjoy

Index

Preface

Cloud Pak for Data is IBM's modern Data and AI platform that includes strategic offerings from its data and AI portfolio delivered in a cloud-native fashion with the flexibility of deployment on any cloud. The platform offers a unique approach to address modern challenges with an integrated mix of proprietary, open source, and third-party services.

You will start with key concepts in modern data management and AI, review real-life use cases, and develop an appreciation of the AI Ladder principle. With this foundation, you will explore how Cloud Pak for Data helps in the elegant implementation of the AI Ladder practice to collect, organize, analyze, and infuse data and trustworthy AI across your business. As you advance, you will also discover the capabilities of the platform and extension services, including how they are packaged and priced. With examples throughout the book, you will gain a deep understanding of the platform, from its rich capabilities and technical architecture to its ecosystem and key go-to-market aspects.

At the end of this IBM book, you will be well-versed in the concepts of IBM Cloud Pak for Data, and be able to apply its prescriptive practices and leverage its capabilities in building a trusted data foundation and accelerate AI adoption in your enterprise.

> **Note**
>
> The content in this book is comprehensive and covers multiple versions in support as of Oct 2021 including version 3.5 and version 4.0. Some of the services, capabilities, and features highlighted in the book might not be relevant to all versions, and as the product evolves we expect a few more changes.
>
> However, the overarching message, value prop, and underlying architecture will remain more or less consistent. Given the rapid progress and product evolution, we decided to be exhaustive while focusing to highlight the core concepts.
>
> We sincerely hope that you will find this book helpful and overlook any inconsistencies attributed to product evolution.

Who this book is for

This book is for business executives, CIOs, CDOs, data scientists, data stewards, data engineers, and developers interested in learning about IBM's Cloud Pak for Data. Knowledge of technical concepts and familiarity with data, analytics, and AI initiatives at various levels of maturity is required to make the most of this book.

What this book covers

Chapter 1, The AI Ladder: IBM's Prescriptive Approach, explores market dynamics, IBM's data and AI portfolio, and a detailed overview of the AI Ladder, what it entails, and how IBM offerings map to the different rungs of the ladder.

Chapter 2, Cloud Pak for Data: A Brief Introduction, covers IBM's modern data and AI platform in detail, along with some of its key differentiators. We will discuss Red Hat OpenShift, the implied cloud benefits it confers, and the platform foundational services that form the basis of Cloud Pak for Data.

Chapter 3, Collect – Making Data Simple and Accessible, offers a flexible approach to address the modern challenges with data-centric delivery, with the proliferation of data both in terms of volume and variety, with a mix of proprietary, open source, and third-party services.

Chapter 4, Organize – Creating a Trusted Analytics Foundation, allows you to learn how Cloud Pak for Data enables Data Ops (data operations), orchestration of people, processes, and technology to deliver trusted, business-ready data to data citizens, operations, applications, and **artificial intelligence** (**AI**) fast.

Chapter 5, Analyzing: Building, Deploying, and Scaling Models with Trust and Transparency, explains how to analyze your data in smarter ways and benefit from visualization and AI models that empower your organization to gain new insights and make better and smarter decisions.

Chapter 6, Multi-Cloud Strategy and Cloud Satellite, offers to operationalize AI throughout the business, allowing your employees to focus on higher-value work.

Chapter 7, IBM and Partner Extension Services, covers the technical concepts underpinning Cloud Pak for Data, including, but not limited to, an architecture overview, common services, Day-2 operations, infrastructure and storage support, and other advanced concepts.

Chapter 8, Customer Use Cases, drills down into the concepts of extension services, how they are packaged and priced, and the various IBM extension services available on Cloud Pak for Data across the **Collect**, **Organize**, **Analyze**, and **Infuse** rungs of the AI ladder.

Chapter 9, Technical Overview, Management, and Administration, addresses the importance of a partner ecosystem, the different tiers of business partners, and how clients can benefit from an open ecosystem on Cloud Pak for Data.

Chapter 10, Security and Compliance, focuses on the importance of business outcomes and key customer use case patterns of Cloud Pak for Data while highlighting the top three use case patterns: data modernization, DataOps, and an automated AI life cycle.

Chapter 11, Storage, looks at how the two critical prerequisites for enterprise adoption, security and governance, are addressed in Cloud Pak for Data.

Chapter 12, Multi-Tenancy, covers the different storage options supported by Cloud Pak for Data and how to configure it for high availability and disaster recovery.

To get the most out of this book

Knowledge of technical concepts and familiarity with data, analytics, and AI initiatives at various levels of maturity is required to make the most of this book.

Software/hardware covered in the book	Operating system requirements
Access to a Cloud Pak for Data installation	Windows, macOS, or Linux
OpenShift or Kubernetes utilities (kubectl and oc)	
The Linux operating system and its security primitives	

If you are using the digital version of this book, we advise you to type the code yourself. Doing so will help you avoid any potential errors related to the copying and pasting of code.

Download the color images

We also provide a PDF file that has color images of the screenshots and diagrams used in this book. You can download it here:

```
https://static.packt-cdn.com/downloads/9781800562127_
ColorImages.pdf
```

Conventions used

There are a number of text conventions used throughout this book.

`Code in text`: Indicates code words in text, database table names, folder names, filenames, file extensions, pathnames, dummy URLs, user input, and Twitter handles. Here is an example: "The Cloud Pak for Data control plane introduces a special persistent volume claim called `user-home-pvc`."

A block of code is set as follows:

```
kubectl get pvc user-home-pvc
NAME             STATUS      VOLUME
CAPACITY    ACCESS MODES    STORAGECLASS    AGE
user-home-pvc    Bound       pvc-44e5a492-9921-41e1-bc42-
b96a9a4dd3dc     10Gi        RWX             nfs-client    33d
```

When we wish to draw your attention to a particular part of a code block, the relevant lines or items are set in bold:

```
Port:              zencoreapi-tls  4444/TCP
TargetPort:        4444/TCP
Endpoints:         10.254.16.52:4444,10.254.20.23:4444
```

Bold: Indicates a new term, an important word, or words that you see on screen. For instance, words in menus or dialog boxes appear in **bold**. Here is an example: "There are essentially two types of host nodes (as presented in the screenshot) – the **Master** and **Compute** (worker) nodes."

> **Tips or important notes**
> Appear like this.

Get in touch

Feedback from our readers is always welcome.

General feedback: If you have questions about any aspect of this book, email us at `customercare@packtpub.com` and mention the book title in the subject of your message.

Errata: Although we have taken every care to ensure the accuracy of our content, mistakes do happen. If you have found a mistake in this book, we would be grateful if you would report this to us. Please visit www.packtpub.com/support/errata and fill in the form.

Piracy: If you come across any illegal copies of our works in any form on the internet, we would be grateful if you would provide us with the location address or website name. Please contact us at copyright@packt.com with a link to the material.

If you are interested in becoming an author: If there is a topic that you have expertise in and you are interested in either writing or contributing to a book, please visit authors.packtpub.com.

Share Your Thoughts

Once you've read *IBM Cloud Pak for Data*, we'd love to hear your thoughts! Scan the QR code below to go straight to the Amazon review page for this book and share your feedback.

https://packt.link/r/1-800-56212-8

Your review is important to us and the tech community and will help us make sure we're delivering excellent quality content.

Section 1: The Basics

In this section, we will learn about market trends, data and AI, IBM's offering portfolio, its prescriptive approach to AI adoption, and an overview of Cloud Pak for Data.

This section comprises the following chapters:

- *Chapter 1, The AI Ladder: IBM's Prescriptive Approach*
- *Chapter 2, Cloud Pak for Data – A Brief Introduction*

1

The AI Ladder – IBM's Prescriptive Approach

Digital transformation is impacting every industry and business, with data and **artificial intelligence** (**AI**) playing a prominent role. For example, some of the largest companies in the world, such as Amazon, Facebook, Uber, and Google, leverage data and AI as a key differentiator. However, not every enterprise is successful in embracing AI and monetizing their data. The AI ladder is IBM's response to this market need – it's a prescriptive approach to AI adoption and entails four simple steps or rungs of the ladder.

In this chapter, you will learn about market dynamics, IBM's Data and AI portfolio, and a detailed overview of the AI ladder. We are also going to cover what it entails and how IBM offerings map to the different rungs of the ladder.

In this chapter, we will be covering the following main topics:

- Market dynamics and IBM's Data and AI portfolio
- Introduction to the AI ladder
- Collect – making data simple and accessible

- Organize – creating a trusted analytics foundation
- Analyze – building and scaling AI with trust and transparency
- Infuse – operationalizing AI throughout the business

Market dynamics and IBM's Data and AI portfolio

The fact is that *every company in the world today is a data company.* As the *Economist* magazine rightly pointed out in 2017, data is the world's most valuable resource and unless you are leveraging your data as a strategic differentiator, you are likely missing out on opportunities.

Simply put, data is the fuel, the cloud is the vehicle, and AI is the destination. The intersection of these *three pillars of IT* is the driving force behind digital transformation disrupting every company and industry. To be successful, companies need to quickly modernize their portfolio and embrace an intentional strategy to re-tool their data, AI, and application workloads by leveraging a cloud-native architecture. So, cloud platforms act as a great enabler by infusing agility, while AI is the ultimate destination, the so-called nirvana that every enterprise seeks to master.

While the benefits of the cloud are becoming obvious by the day, there are still several enterprises that are reluctant to embrace the public cloud right away. These enterprises are, in some cases, constrained by regulatory concerns, which make it a challenge to operate on public clouds. However, this doesn't mean that they don't see the value of the cloud and the benefits derived from embracing the cloud architecture. Everyone understands that the cloud is the ultimate destination, and taking the necessary steps to prepare and modernize their workloads is not an option, but a survival necessity:

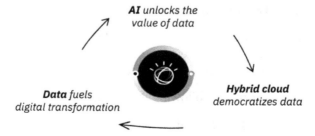

Figure 1.1 – What's reshaping how businesses operate? The driving forces behind digital transformation

IBM enjoys a strong Data and AI portfolio, with 100+ products being developed and acquired over the past 40 years, including some marquee offerings such as Db2, Informix, DataStage, Cognos Analytics, SPSS Modeler, Planning Analytics, and more. The depth and breadth of IBM's portfolio is what makes it stand out in the market. With Cloud Pak for Data, IBM is doubling down on this differentiation, further simplifying and modernizing its portfolio as customers look to a hybrid, multi-cloud future.

Introduction to the AI ladder

We all know data is the foundation for businesses to drive smarter decisions. Data is what fuels digital transformation. But it is AI that unlocks the value of that data, which is why AI is poised to transform businesses with the potential to add almost 16 trillion dollars to the global economy by 2030. You can find the relevant source here: `https://www.pwc.com/gx/en/issues/data-and-analytics/publications/artificial-intelligence-study.html`.

However, the adoption of AI has been slower than anticipated. This is because many enterprises do not make a conscious effort to lay the necessary data foundation and invest in nurturing talent and business processes that are critical for success. For example, the vast majority of AI failures are due to data preparation and organization, not the AI models themselves. Success with AI models is dependent on achieving success in terms of how you collect and organize data. Business leaders not only need to understand the power of AI but also how they can fully unleash its potential and operate in a hybrid, multi-cloud world.

This section aims to demystify AI, common AI challenges and failures, and provide a unified, prescriptive approach (which we call "the AI ladder") to help organizations unlock the value of their data and accelerate their journey to AI.

As companies look to harness the potential of AI and identify the best ways to leverage data for business insights, they need to ensure that they start with a clearly defined business problem. In addition, you need to use data from diverse sources, support best-in-class tools and frameworks, and run models across a variety of environments.

According to a study by *MIT Sloan Management Review*, 81% of business leaders (`http://marketing.mitsmr.com/offers/AI2017/59181-MITSMR-BCG-Report-2017.pdf`) do not understand the data and infrastructure required for AI and "*No amount of AI algorithmic sophistication will overcome a lack of data [architecture] – bad data is simply paralyzing.*"

Put simply: *There is no AI without IA* (information architecture).

IBM recognizes this challenge our clients are facing. As a result, IBM built a prescriptive approach (known as **the AI ladder**) to help clients with the aforementioned challenges and accelerate their journey to AI, no matter where they are on their journey. It allows them to simplify and automate how organizations turn data into insights by unifying the collection, organization, and analysis of data, regardless of where it lives. By climbing the AI ladder, enterprises can build a governed, efficient, agile, and future-proof approach to AI. Furthermore, it is also an organizing construct that underpins the Data and AI product portfolio of IBM.

It is critical to remember that AI is not magic and requires a thoughtful and well-architected approach. Every step of the ladder is critical to being successful with AI.

The rungs of the AI ladder

The following diagram illustrates IBM's prescriptive approach, also known as **the AI ladder**:

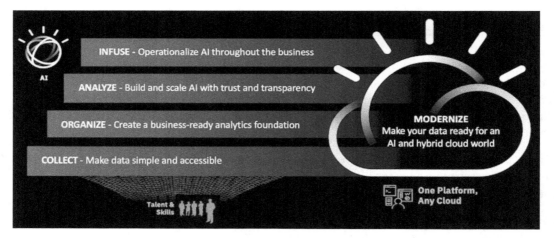

Figure 1.2 – The AI ladder – a prescriptive approach to the journey of AI

The AI ladder has four steps (often referred to as the *rungs* of the ladder). They are as follows:

1. **Collect: Make data simple and accessible**. Collect data of every type regardless of where it lives, enabling flexibility in the face of ever-changing data sources.

2. **Organize: Create a business-ready analytics foundation**. Organize all the client's data into a trusted, business-ready foundation with built-in governance, quality, protection, and compliance.

3. **Analyze: Build and scale AI with trust and explainability**. Analyze the client's data in smarter ways and benefit from AI models that empower the client's team to gain new insights and make better, smarter decisions.

4. **Infuse: Operationalize AI throughout the business**. You should do this across multiple departments and within various processes by drawing on predictions, automation, and optimization. Craft an effective AI strategy to realize your AI business objectives. Apply AI to automate and optimize existing workflows in your business, allowing your employees to focus on higher-value work.

Spanning the four steps of the AI ladder is the concept of **Modernize** from IBM, which allows clients to simplify and automate how they turn data into insights. It unifies collecting, organizing, and analyzing data within a multi-cloud data platform known as Cloud Pak for Data.

IBM's approach starts with a simple idea: run anywhere. This is because the platform can be deployed on the customer's infrastructure of choice. IBM supports Cloud Pak for Data deployments on every major cloud platform, including Google, Azure, AWS, and IBM Cloud. You can also deploy Cloud Pak for Data platforms on-premises in your data center, which is extremely relevant for customers who are focused on a hybrid cloud strategy.

The way IBM supports Cloud Pak for Data on all these infrastructures is by layering **Red Hat OpenShift** at its core. This is one of the key reasons behind IBM's acquisition of Red Hat in 2019. The intention is to offer customers the flexibility to scale across any infrastructure using the world's leading open source steward: Red Hat. OpenShift is a Kubernetes-based platform that also allows IBM to deploy all our products through a modern container-based model. In essence, all the capabilities are rearchitected as microservices so that they can be provisioned as needed based on your enterprise needs.

Now that we have introduced the concept of the AI ladder and IBM's Cloud Pak for Data platform, let's spend some time focusing on the individual rungs of the AI ladder and IBM's capabilities that make it stand out.

Collect – making data simple and accessible

The Collect layer is about putting your data in the appropriate persistence store to efficiently collect and access all your data assets. A well-architected "Collect" rung allows an organization to leverage the appropriate data store based on the use case and user persona; whether it's Hadoop for data exploration with data scientists, OLAP for delivering operational reports leveraging business intelligence or other enterprise visualization tools, NoSQL databases such as MongoDB for rapid application development, or some mixture of them all, you have the flexibility to deliver this in a single, integrated manner with the Common SQL Engine.

IBM offers some of the best database technology in the world for addressing every type of data workload, from **Online Transactional Processing (OLTP)** to **Online Analytical Processing (OLAP)** to Hadoop to fast data. This allows customers to quickly change as their business and application needs change. Furthermore, IBM layers a Common SQL Engine across all its persistence stores to be able to write SQL once, and leverage your persistence store of choice, regardless of whether it is IBM Db2 or open source persistence stores such as MongoDB or Hadoop. This allows for portable applications and saves enterprises significant time and money that would typically be spent on rewriting queries for different flavors of persistence. Also, this enables a better experience for end users and a faster time to value.

IBM's Db2 technology is enabled for natural language queries, which allows non-SQL users to search through their OLTP store using natural language. Also, Db2 supports **Augmented Data Exploration (ADE)**, which allows users to access the database and visualize their datasets through automation (as opposed to querying data using SQL).

To summarize, Collect is all about collecting data to capture newly created data of all types, and then bringing it together across various silos and locations to make it accessible for further use (up the AI ladder). In IBM, the Collect rung of the AI ladder is characterized by three key attributes:

- **Empower**: IT architects and developers in enterprises are empowered as they are offered a complete set of fit-for-purpose data capabilities that can handle all types of workloads in a self-service manner. This covers all workloads and data types, be it structured or unstructured, open source or proprietary, on-premises or in the cloud. It's a single portfolio that covers all your data needs.

- **Simplify**: One of the key tenets of simplicity is enabling self-service, and this is realized rather quickly in a containerized platform built using cloud-native principles. For one, provisioning new data stores involves a simple click of a button. In-place upgrades equate to zero downtime, and scaling up and down is a breeze, ensuring that enterprises can quickly react to business needs in a matter of minutes as opposed to waiting for weeks or months. Last but not least, IBM is infusing AI into its data stores to enable augmented data exploration and other automation processes.

- **Integrate**: Focuses on the need to make data accessible and integrate well with the other rungs of the AI ladder. Data virtualization, in conjunction with data governance, enables customers to access a multitude of datasets in a single view, with a consistent glossary of business terms and associated lineage, all at your fingertips. This enables the democratization of enterprise data accelerating AI initiatives and driving automation to your business. The following diagram summarizes the key facets of the Collect rung of the AI ladder:

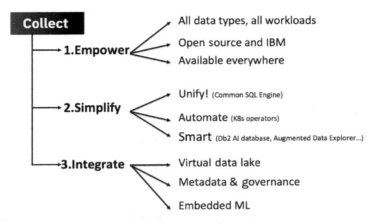

Figure 1.3 – Collect – making data simple and accessible

Our portfolio of capabilities, all of which support the Collect rung, can be categorized into four workload domains in the market:

1. First, there's the traditional **operational database**. This is your system of records, your point of sales, and your transactional database.

2. **Analytics databases** are in high demand as the amount of data is exploding. Everyone is looking for new ways to analyze data at scale quickly, all the way from traditional reporting to preparing data for training and scoring AI models.

3. **Big data**. The history of having a data lake using Hadoop at petabyte scale is now slowly transforming into the separation of storage and compute, with Cloud Object Storage and Spark playing key roles. The market demand for data lakes is clearly on an upward trajectory.

4. Finally, **IoT** is quickly transforming several industries, and the fast data area is becoming an area of interest. This is the market of the future, and IBM is addressing requirements in this space through real-time data analysis.

Next, we will explore the importance of organizing data and what it entails.

Organize – creating a trusted analytics foundation

Given that data sits at the heart of AI, organizations will need to focus on the quality and governance of their data, ensuring it's accurate, consistent, and trusted. However, many organizations struggle to streamline their operating model when it comes to developing data pipelines and flows.

Some of the most common data challenges include the following:

- Lack of data quality, governance, and lineage
- Trustworthiness of structured and unstructured data
- Searchability and discovery of relevant data
- Siloed data across the organization
- Slower time-to-insight for issues that should be real time-based
- Compliance, privacy, and regulatory pressures
- Providing self-service access to data

To address these many data challenges, organizations are transforming their approach to data: they are undergoing application modernization and refining their data strategies to stay compliant while still fueling innovation.

Delivering trusted data throughout your organization requires the adoption of new methodologies and automation technologies to drive operational excellence in your data operations. This is known as DataOps. This is also referred to as "enterprise data fabric" by many and plays a critical role in ensuring that enterprises are gaining value from their data.

DataOps corresponds to the *Organize* rung of IBM's AI ladder; it helps answer questions such as the following:

- What data does your enterprise have, and who owns it?
- Where is that data located?
- What systems are using the data in question and for what purposes?
- Does the data meet all regulatory and compliance requirements?

DataOps also introduces agile development processes into data analytics so that data citizens and business users can work together more efficiently and effectively, resulting in a collaborative data management practice. And by using the power of automation, DataOps helps solve the issues associated with inefficiencies in data management, such as accessing, onboarding, preparing, integrating, and making data available.

DataOps is defined as the orchestration of people, processes, and technology to deliver trusted, high-quality data to whoever needs it.

People empowering your data citizens

A modern enterprise consists of many different "data citizens" – from the chief data officer; to data scientists, analysts, architects, and engineers; to the individual line of business users who need insights from their data. The Organize rung is about creating and sustaining a data-driven culture that enables collaboration across an organization to drive agility and scale.

Each organization has unique requirements where stakeholders in IT, data science, and the business lines need to add value to drive a successful business. Also, because governance is one of the driving forces needed to support DataOps, organizations can leverage existing data governance committees and lessons from tenured data governance programs to help establish this culture and commitment.

The benefits of DataOps mean that businesses function more efficiently once they implement the right technology and develop self-service data capabilities that make high-quality, trusted data available to the right people and processes as quickly as possible. The following diagram shows what a DataOps workflow might look like: architects, engineers, and analysts collaborate on infrastructure and raw data profiling; analysts, engineers, and scientists collaborate on building analytics models (whether those models use AI); and architects work with business users to operationalize those models, govern the data, and deliver insights to the points where they're needed.

Individuals within each role are designated as **data stewards** for a particular subset of data. The point data citizens of the DataOps methodology is that each of these different roles can rely on seeing data that is accurate, comprehensive, secure, and governed:

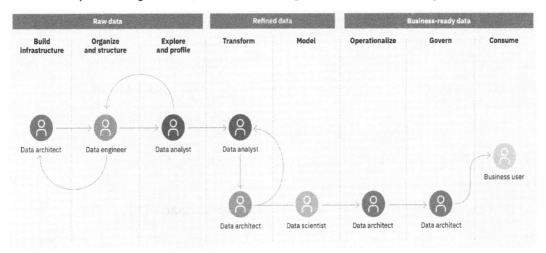

Figure 1.4 – DataOps workflow by roles

IBM has a rich portfolio of offerings (now available as services within Cloud Pak for Data) that address all the different requirements of DataOps, including data governance, automated data discovery, centralized data catalogs, ETL, governed data virtualization, data privacy/masking, master data management, and reference data management.

Analyze – building and scaling models with trust and transparency

Enterprises are either building AI or buying AI solutions to address specific requirements. In the case of a build scenario, companies would benefit significantly from commercially available data science tools such as Watson Studio. IBM's Watson Studio not only allows you to make significant productivity gains but also ensures collaboration among the different data scientists and user personas.

Investing in building AI and retraining employees can have a significant payoff. Pioneers across multiple industries are building AI and separating themselves from laggards:

- In **construction**, they're using AI to optimize infrastructure design and customization.

- In **healthcare**, companies are using AI to predict health problems and disease symptoms.

- In **life science**, organizations are advancing image analysis to research drug effects.

- In **financial services**, companies are using AI to assist in fraud analysis and investigation.

- Finally, **autonomous vehicles** are using AI to adapt to changing conditions in vehicles, while call centers are using AI for automating customer service.

However, several hurdles remain, and enterprises face significant challenges in operationalizing AI value.

There are three areas that we need to tackle:

- **Data**: 80% of time is spent preparing data versus building AI models.

- **Talent**: 65% find it difficult to fund or acquire AI skills.

- **Trust**: 44% say it's very challenging to build trust in AI outcomes.

Source: *2019 Forrester, Challenges That Hold Firms Back From Achieving AI Aspirations.*

Also, it's worth pointing out that building AI models is the easy part. The real challenge lies in deploying those AI models into production, monitoring them for accuracy and drift detection, and ensuring that this becomes the norm.

IBM's AI tools and runtimes on Cloud Pak for Data present a differentiated and extremely strong set of capabilities. Supported by the Red Hat OpenShift and Cloud Pak for Data strategy, IBM is in a position to set and lead the market for AI tools. There are plenty of point AI solutions from niche vendors in the market, as evidenced from the numerous analyst reports; however, none of them are solving the problem of putting AI into production in a satisfactory manner. The differentiation that IBM brings to the market is the full end-to-end AI life cycle:

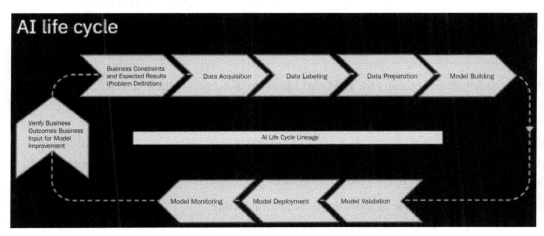

Figure 1.5 – AI life cycle

Customers are looking for an integrated platform for a few reasons. Before we get to these reasons, the following teams care about the integrated platform:

1. Data science teams are looking for integrated systems to manage assets across the AI life cycle and across project team members.

2. **Chief Data Officer** are looking to govern AI models and the data associated with them. **Chief Risk Officer (CRO)** are looking to control the risks that these models expose by being integrated with business processes.

3. Extended AI application teams need integration so that they can build, deploy, and run seamlessly. In some situations, **Chief information officer (CIOs)**/business technology teams who want to de-risk and reduce the costs of taking an AI application to production are responsible for delivering a platform.

> **Customer Use Case**
>
> A Fortune 500 US bank is looking for a solution in order to rapidly deploy machine learning projects to production. The first step in this effort is to put in place a mechanism that allows project teams to deliver pilots without having to go through full risk management processes (from corporate risk/**MRM** teams). They call this *a soft launch*, which will work with some production data. The timeline to roll out projects is 6-9 months from conceptualization to pilot completion. This requirement is being championed (and will need to be delivered across the bank) by the business technology team (who are responsible for the *AI operations portal*). The idea is that this will take the load away from MRM folks who have too much on their plate but still have a clear view of how and what risk was evaluated. **LOB** will be using the solution every week to retrain models. However, before that, they will upload a CSV file, check any real-time responses, and pump data to verify that the model is meeting strategy goals. All this must be auto-documented.

One of the key differentiators for IBM's AI life cycle is AutoAI, which allows data scientists to create multiple AI models and score them for accuracy. Some of these tests are not supposed to be black and white.

Several customers are beginning to automate AI development. Due to this, the following question arises: why automate model development? Because if you can automate the AI life cycle, you can enhance your success rate.

An automated AI life cycle allows you to do the following:

- **Expand your talent pool**: This lowers the skills required to build and operationalize AI models
- **Speed up time to delivery**: This is done by minimizing mundane tasks.
- **Increase the readiness of AI-powered apps**: This is done by optimizing model accuracy and KPIs.
- **Deliver real-time governance**: This improves trust and transparency by ensuring model management, governance, explainability, and versioning.

Next, we will explore how AI is operationalized in enterprises to address specific use cases and drive business value.

Infuse – operationalizing AI throughout the business

Building insights and AI models is a great first step, but unless you infuse them into your business processes and optimize outcomes, AI is just another fancy technology. Companies who have automated their business processes based on data-driven insights have disrupted the ones who haven't – case in point being Amazon in retail, who has upended many traditional retailers by leveraging data, analytics, and AI to streamline operations and gain a leg up on the competition. The key here is to marry technology with culture and ensure that employees are embracing AI and infusing it into their daily decision making:

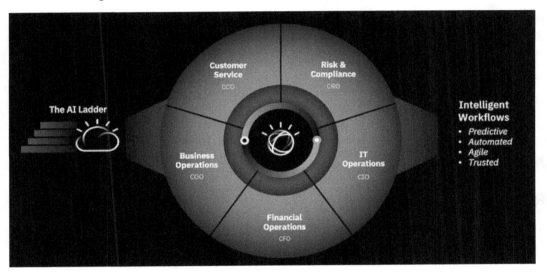

Figure 1.6 – Infuse – AI is transforming how businesses operate

The following are some diverse examples of companies infusing AI into their business processes. These are organized along five key themes:

- **Customer service (business** owner: **CCO**): Customer care automation, Customer 360, customer data platform.

- **Risk and compliance (business** owner: **CRO**): Governance risk and compliance.

- **IT operations (business** owner: **CIO**): Automate and optimize IT operations.

- **Financial operations (business** owner: **CFO**): Budget and optimize across multiple dimensions.

- **Business operations (business** owner: **COO**): Supply chain, human resources management.

Customer service

Customer service is changing by the day with automation driven by chatbots and a 360-degree view of the customer becoming more critical. While there is an active ongoing investment on multiple fronts within IBM, the one that stands out is IBM's *Watson Anywhere* campaign, which allows customers to buy Cloud Pak for Data Watson services (Assistant, Discovery, and API Kit) at a discount and have it deployed.

> **Customer Use Case**
>
> A technology company that offers mobile, telecom, and CRM solutions is seeing a significant demand for intelligent call centers and invests in an AI voice assistant on IBM Cloud Pak for Data. The objective is to address customers' queries automatically, reducing the need for human agents. Any human interaction happens only when detailed consultation is required. This frees up call center employees to focus on more complex queries as opposed to handling repetitive tasks, thus improving the overall operational efficiency and quality of customer service, not to mention reduced overhead costs. This makes building intelligent call centers simpler, faster, and more cost-effective to operate. Among other technologies, that proposed solution uses Watson Speech to Text, which converts voice into text to help us understand the context of the question. This allows AI voice agents to quickly provide the best answer in the context of a customer inquiry.

Risk and compliance

Risk and compliance is a broad topic and companies are struggling to ensure compliance across their processes. In addition to governance risk and compliance, you also need to be concerned about the financial risks posed to big banks. IBM offers a broad set of out-of-the-box solutions such as OpenPages, Watson Financial Crimes Insight, and more, which, when combined with AI governance, deliver significant value, not just in addressing regulatory challenges, but also in accelerating AI adoption.

IT operations

With IT infrastructure continuing to grow exponentially, there is no reason to believe that it'll decline any time soon. On the contrary, the complexity of operating IT infrastructure is not a simple task and requires the use of AI to automate operations and proactively identify potential risks. Mining data to predict and optimize operations is one of the key use cases of AI. IBM has a solution called **Watson AIOps** on the Cloud Pak for Data platform, which is purpose-built to address this specific use case.

Financial operations

Budgeting and forecasting typically involves several stakeholders collaborating across the enterprise to arrive at a steady answer. However, this requires more than hand waving. IBM's Planning Analytics solution on Cloud Pak for Data is a planning, budgeting, forecasting, and analysis solution that helps organizations automate manual, spreadsheet-based processes and link financial plans to operational tactics.

IBM Planning enables users to discover insights automatically, directly from their data, and drive decision making with the predictive capabilities of IBM Watson. It also incorporates scorecards and dashboards to monitor KPIs and communicate business results through a variety of visualizations.

Business operations

Business operations entails several domains, including supply chain management, inventory optimization, human resources management, asset management, and more. Insights and AI models developed using the Cloud Pak for Data platform can be leveraged easily across their respective domains. There are several examples of customers using IBM solutions.

> **Customer Use Case**
>
> *A well-known North American healthcare company* was trying to address a unique challenge. They used AI to proactively identify and prioritize at-risk sepsis patients. This required an integrated platform that could manage data across different silos to build, deploy, and manage AI models at scale while ensuring trust and governance. With Cloud Pak for Data, the company was able to build a solution in 6 weeks, which would typically take them 12 months. This delivered projected cost savings of ~$48 K per patient, which is a significant value.

Digital transformation is disrupting our global economy and will bring in big changes in how we live, learn, work, and entertain; and in many cases, this will accelerate the trends we've been seeing across industries. This also applies to how data, analytics, and AI workloads will be managed going forward. Enterprises taking the initiative and leveraging the opportunity to streamline, consolidate, and transform their architecture will come out ahead both in sustaining short-term disruptions and in modernizing for an evolving and agile future. IBM's prescriptive approach to the AI ladder is rooted in a simple but powerful belief that having a strong information architecture is critical for successful AI adoption. It offers enterprises an organizational structure to adopt AI at scale.

The case for a data and AI platform

In the previous section, we introduced IBM's prescriptive approach to operationalizing AI in your enterprise, starting with making data access simple and organizing data into a trusted foundation for advanced analytics and AI. Now, let's look at how to make that a reality.

From an implementation perspective, there are many existing and established products in the industry, some from IBM or its partners and some from competitors. However, there is a significant overhead to making all these existing products work together seamlessly. It also gets difficult when you look at hybrid cloud situations. Not every enterprise has the IT expertise to integrate disparate systems to enable their end users to collaborate and deliver value quickly. Reliably operating such disparate systems securely also stretches IT budgets and introduces so much complexity that it distracts the enterprise from achieving their business goals in a timely and cost-efficient fashion.

Due to this, there is a need for a data and AI platform that provides a standardized technology stack and ensures simplicity in operations. The following diagram shows what is typically needed in an enterprise to enable AI and get full value from their data. It also shows the different personas in a typical enterprise, with different roles and responsibilities and at different skill levels, all of which need to tightly collaborate to deliver on the promise of trusted AI:

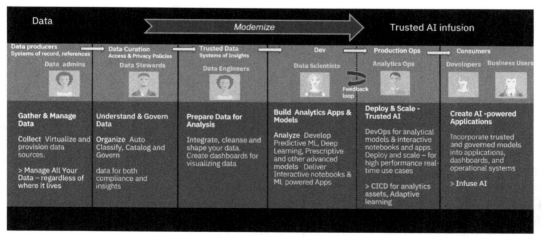

Figure 1.7 – Silos hinder operationalizing AI in the enterprise

Most enterprises would have their data spread across multiple data sources, some on-premises and some in the public cloud and typically in different formats. How would the enterprise make all their valuable data available for their data scientists and analysts in a secure manner? If data science tooling and frameworks cannot work with data in place, at scale, they may even need to build additional expensive data integration and transformation pipelines, as well as storing data in new data warehouses or lakes. If a steward is unable to easily define data access policies across all data in use, ensuring that sensitive data is masked or obfuscated, from a security and compliance perspective, it will become unsafe (or even illegal) to make data available for advanced analytics and AI.

The following diagram expands on the need for different systems to integrate, and users to collaborate closely. It starts with leaders setting expectations on the business problems to address using AI techniques. Note that these systems and tasks span both the development and operational production aspects of the enterprise:

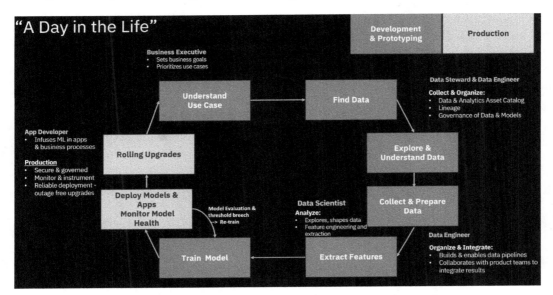

Figure 1.8 – Typical cross-system tasks and cross-persona interactions

This flow is also circular since the systems need to account for feedback and have a clear way of measuring whether the implementation has met the objectives stated. If the consumers of the data do not have visibility into the quality of data or can't ensure the data is not stale, any AI that's built from such data would always be suspect and would therefore pose an increased risk. Fundamentally, the absence of a foundational infrastructure that ties all these systems together can make implementing the AI ladder practice complicated, or in some cases, impossible. What is needed is a reliable, scalable, and modern **data and AI platform** that can break down these silos, easily integrate systems, and enable collaboration between different user personas, even via a single integrated experience.

Summary

In this chapter, you learned about market dynamics and IBM's Data and AI portfolio, were provided with an overview of IBM's prescriptive approach to AI, known as *the AI ladder*, and learned how IBM offerings map to the different rungs of the ladder. In the next chapter, we will provide a thorough introduction to Cloud Pak for Data, IBM's Data and AI platform that enables enterprises to implement *the AI ladder* anywhere in a modular fashion while leveraging modern cloud-native frameworks.

2
Cloud Pak for Data: A Brief Introduction

Cloud Pak for Data was launched in May 2018. It is a platform for IBM's modern data and AI that provides a unified experience and automation as its core tenants. It includes all of IBM's strategic offerings from its data and AI portfolio, delivering cloud-native benefits and the flexibility to deploy on any cloud.

In this chapter, you will learn about the platform in detail along with some of its key differentiators. We will discuss Red Hat OpenShift, the implied cloud benefits it confers, and the platform foundational services that form the basis of Cloud Pak for Data.

The following topics will be covered in this chapter:

- Cloud Pak for Data overview
- Unique differentiators, key use cases, and customer adoption
- Product details
- Red Hat OpenShift

The case of a data and AI platform – recap

In the previous chapter, we introduced IBM's prescriptive approach to operationalizing data and AI in your enterprise, and subsequently, we introduced you to the broad capabilities of Cloud Pak for Data. In this section, we will outline how Cloud Pak for Data powers the implementation of the AI Ladder practice in your enterprise.

Cloud Pak for Data, as an open platform, breaks down silos between different products, enabling them to be manifested as cloud-native services on top of the same OpenShift Kubernetes platform. Cloud Pak for Data enables such services to easily integrate through a common control plane and, importantly, provides a personalized, integrated user experience for end users to collaborate naturally, without even being aware of product or service boundaries.

On this one platform, capabilities such as data virtualization help modernize data access for the hybrid cloud and data can be easily accessed in place across clouds. Data cataloging and governance services can piggyback on data virtualization and other data sources to deliver on broader governed data access plane solutions, which in turn facilitates trust in data that is made available to data science, analytics reporting, and machine learning services. On the same platform, operations teams can deploy ML models at scale, enabling model monitoring with bias detection to support continuous model evolutions.

Development, experimentation, and production deployments can be supported on the same platform, promoting collaborations between teams of users with varying roles and responsibilities to work closely. Authentication and authorization, customized in the control plane to comply with enterprise regulations, enable seamless access across all services. Consistent APIs and concepts such as user groups, API keys, and tokens enable the development of custom applications and solutions in a standard manner.

Administration and operations are simplified through operators powered by Red Hat OpenShift Kubernetes. Portability of deployment across on-premises and public cloud infrastructure is assured, enabling bursts into the cloud when needed. Cloud-native practices enable continuous delivery and integration for enterprise applications and solutions. Security controls and operational auditing are built into the platform to ensure adherence to compliance requirements in the enterprise. The deployment and management of multiple tenants on the same set of shared compute resources also make Cloud Pak for Data much more cost-effective for an enterprise.

Overview of Cloud Pak for Data

In *Chapter 1, The AI Ladder: IBM's Prescriptive Approach*, you were introduced to IBM's prescriptive methodology to AI adoption, **the AI ladder**, and how enterprises can embrace this framework through IBM's modern multi-cloud data and AI platform, Cloud Pak for Data. The concept of an integrated platform to manage data, analytics, and AI services is not new – there has always been a demand for an end-to-end offering spanning the different rungs of the AI ladder. Enterprises addressed this need through custom integrations by often tapping into global consulting organizations for resources and expertise.

With the advent of public cloud and service-oriented architectures, it is now becoming commonplace to procure pre-integrated software platforms. These platforms minimize the integration effort while still providing the flexibility to meet specific business requirements.

Cloud Pak for Data was launched on May 29, 2018, leveraging Kubernetes technology, which allowed the platform to be deployed on any public cloud and/or private cloud in enterprise data centers. This enabled clients to leverage cloud-native benefits such as easy provisioning, scaling, simplified management, and seamless upgrades, while sticking to their infrastructure of choice: the *private cloud* or any *public cloud* such as AWS, Azure, or IBM Cloud. It's also worth highlighting that the data and AI services underpinning the platform have been tried and tested by thousands of enterprise customers over the past few decades. So, while the integrated platform is new and modern, the capabilities and functionality behind it have been proven at scale. This is critical for enterprise customers who cannot afford to make risky investments as they are looking to modernize their data and AI workloads. The following diagram illustrates the various releases of Cloud Pak for Data since its launch:

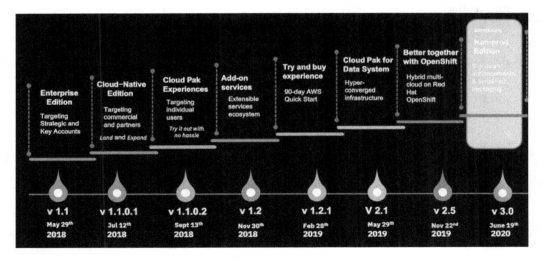

Figure 2.1 – IBM Cloud Pak for Data journey: 8 releases since launch

Now, let's explore the key differentiators of Cloud Pak for Data and the typical use case patterns that enterprises use it for.

Exploring unique differentiators, key use cases, and customer adoption

Cloud Pak for Data offers clients a single platform with a rich ecosystem of modular data and AI services that provides simplicity, fast adaptability to change, and reduced reliance on specialization and hard-to-find skills. Unlike the typical mix of many isolated tools for databases, data operations, data science, data analytics, and visualization we see most of our customers/enterprises using, Cloud Pak for Data is one common operational fabric for all data and AI services. The result is a delightful, unified experience, cost reduction from the automation of manual mundane tasks, and enforced data governance and security.

Cloud Pak for Data is the new way to solve complex business problems with a hybrid-cloud data and AI platform that runs anywhere.

Here are the unique differentiators of the platform:

- **A unified experience**: The end-to-end integrated platform amplifies productivity while enabling collaboration across different personas and functional divisions.

- **Automation**: AI is leveraged to automate a number of tasks such as building models (Auto AI), discovering and cataloging datasets (data discovery), and many more.

- **Governance and Security**: This is infused throughout the platform and is a major differentiator. While data governance enables customers to ensure compliance and self-service analytics, AI governance infuses trust and drives the adoption of AI projects within enterprises while also addressing the needs of regulatory compliance.

- **Deploy anywhere**: Cloud Pak for Data runs on any public or private cloud and connects data everywhere. This provides enterprises with flexibility and a choice of infrastructure and avoids vendor lock-in (*the independence to migrate workloads to a different cloud provider if needed down the road*).

Let's now explore a few key use cases of Cloud Pak for Data.

Key use cases

The following are some of the key use cases of Cloud Pak for Data:

- **AI life cycle**: Data science teams are looking for integrated systems to manage assets across the AI lifecycle, while enterprise **Chief Data Officers** (**CDOs**) want to ensure the governance of AI models and datasets associated with them. Cloud Pak for Data enables the end-to-end AI lifecycle all the way from data preparation to building and deploying models, to managing them at scale, all the while ensuring the governance of data and AI models. Enterprises can realize an integrated experience, automation, and the benefits of AI governance including bias detection, explainability, model fairness, and drift detection:

"AI Ladder" Themes	Organize
	Analyze
"Cloud Pak for Data" services	Watson Knowledge Catalog
	Watson Studio
	Watson Machine Learning
	Watson OpenScale

Figure 2.2 – AI lifecycle

- **Data modernization**: Enterprises are grappling with the proliferation of data, with data distributed across multiple silos, databases, and clouds. Cloud Pak for Data addresses this through governed data virtualization enabling self-service access to data in real time. It also includes a comprehensive set of capabilities to discover, prep, transform, govern, catalog, and access data at scale across all enterprise data sources.

"AI Ladder" Themes	Collect
	Organize
"Cloud Pak for Data" services	Watson Knowledge Catalog
	Data Virtualization
	Big SQL (optional)
	Data Stores – Db2, MongoDB, PostgreSQL (Optional)

Figure 2.3 – Data modernization

- **Data ops**: Organizations want to enforce policy controls to allow access to *all* relevant datasets. Cloud Pak for Data enables enterprises to scale their data operations and enforce policies on individual columns and rows so that assets with sensitive data can still be used. Among other things, this pattern includes capabilities around data preparation, data quality, lineage, and regulatory compliance:

"AI Ladder" Themes	Organize
"Cloud Pak for Data" services	Watson Knowledge Catalog Data Refinery Data Stage (Optional) IBM Regulatory Accelerator (Optional)

Figure 2.4 – Data ops

- **Self-service analytics**: Data is the world's most valuable asset and is a competitive differentiator if used appropriately. Data warehouses and BI is one of the foundational use cases, which entails collecting relevant data and building reports/dashboards to derive business insights. Cloud Pak for Data includes everything you need to persist, visualize, and analyze data at scale:

"AI Ladder" Themes	Collect Infuse
"Cloud Pak for Data" services	Db2 Warehouse Cognos Dashboards Cognos Analytics (Optional)

Figure 2.5 – Self-service analytics

- **AI for financial operations**: Financial and operation planning in enterprises is often multi-dimensional, spanning a number of people and departments. Cloud Pak for Data enables automated and integrated planning across your organization, from financial planning and analysis to workforce planning, and sales forecasting to supply chain planning. Empower your organization to deliver more agile, reliable plans and forecasts to drive better business performance:

"AI Ladder" Themes	Infuse
"Cloud Pak for Data" services	Planning Analytics Cognos Analytics (Optional)

Figure 2.6 – AI for financial operations

- **AI for customer care**: Automating customer care is one of the key enterprise use cases of AI. Among other things, Watson helps reduce the time to resolution and call volumes and increases customer satisfaction. **Watson Assistant (WA)** can provide AI-powered automated assistance to customers or employees through web/mobile or voice channels and enable human agents to better handle customer inquiries. **Watson Discovery (WD)** compliments Watson Assistant and can help unlock insights from complex business content – manuals, contracts, and scanned PDFs:

"AI Ladder" Themes	*Infuse*
"Cloud Pak for Data" services	*Watson Assistant* *Watson Discovery (Optional)*

Figure 2.7 – AI for customer care

Customer use case: AI claim processing

A European government agency has a workforce consisting of thousands of employees directly involved in benefits claim processing. The workforce is aging and retiring, leaving the client short on skilled employees who can process benefit claims.

Because of this, the agency decided to create a benefits platform and try out claims processing using an AI solution based on IBM Cloud Pak for Data and other products. Given that this is a government agency, the solution needs to be on-premises, constraining the available options.

The AI lifecycle services within Cloud Pak for Data (Watson Studio and Watson Machine Learning) are leveraged to create machine learning models for the benefits claims process. The benefits of the engagement include cost savings, speed, and quality in the benefits claims process.

Customer use case: data and AI platform

A fortune 500 financial services company had three major issues slowing their journey to become a data-driven company. Their data was all over their enterprise—some in Hadoop, some in SQL, some in Excel spreadsheets—and impossible to access easily. They didn't have an enterprise data governance strategy so there was **no unified set of definitions**, and no one knew the health of the data or its provenance.

Lastly, every data science project they tried took an unacceptably long time to run since they were using local machines, limited by consumer-grade hardware. The client wanted to do the following:

- Create a single data fabric layer
- Establish an enterprise metadata layer
- Perform data science projects

Cloud Pak for Data offers an **end-to-end platform** that addresses all the above requirements based on an open ecosystem and advanced AI features. Clients can start to see some major benefits across their organization. The out-of-the-box metadata and data governance capabilities will enable the collaboration and unification of processes across the enterprise, saving months of work.

Cloud Pak for Data: additional details

When clients start with IBM Cloud Pak for Data, they get the foundational capabilities for the *Collect*, *Organize*, and *Analyze* rungs on the AI Ladder, all integrated via the flexibility of cloud-native microservices. They can access data, create policy-based catalogs, build open source AI/ML models, visualize data with dashboards, and more.

You can think of these capabilities as cloud-native microservice versions of IBM's core offerings (such as Db2, Infosphere, DataStage, Watson Studio, and Cognos Analytics), integrated into a platform – no assembly required. We can see the detailed structure here:

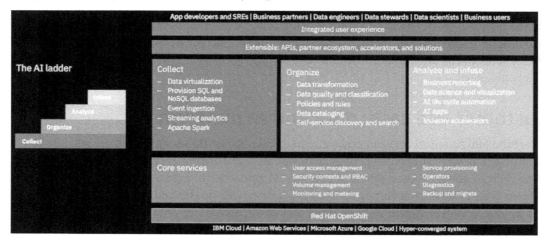

Figure 2.8 – Cloud Pak for Data: unified, modular, and deployable anywhere

Cloud Pak for Data is an open platform with a healthy mix of proprietary, third-party, and open source content, which we will explore next.

An open ecosystem

Cloud Pak for Data is an open and modular platform that allows clients to deploy services they need. One of the highlights of the platform is its vibrant ecosystem, which consists of IBM services (further segmented into base and premium), third-party services, industry accelerators, and open source content.

The following services constitute the base Cloud Pak for Data platform:

- **Db2 Warehouse**: A data warehouse designed for high performance and in-database analytics. Runs on a single node for cost-efficiency or on multiple nodes for improved performance.

- **Data Virtualization**: Built leveraging IBM's patented technology, Data Virtualization enables access to distributed data from disparate data sources in real time without physically moving the data. Furthermore, queries are optimized to run where data resides, cutting down the response time.

- **Db2 EventStore**: A datastore designed to rapidly ingest and analyze streamed data for event-driven applications.

- **Streams**: Build solutions that drive real-time business decisions by combining streaming and stored data with analytics.

- **Watson Knowledge Catalog**: Find the right data fast. Discover relevant, curated data assets using intelligent recommendations and user reviews.

- **IBM Regulatory Accelerator**: Streamline the process of complying with regulations.

- **Watson Studio**: Has three key components as seen here:

  Build: Build custom models and infuse your business with AI and machine learning. This component covers the model development capabilities in Watson Studio (with AutoAI as a key differentiator).

  Deploy: *The model deployment capabilities in Watson Studio* (previously Watson Machine Learning). Deploy machine-learning models in production at scale.

  Trust: The trusted AI capabilities in Watson Studio, including explainability, bias, and drift detection (previously Watson OpenScale). Infuse your AI with trust and transparency and understand how your AI models make decisions to detect and mitigate bias.

- **SPSS Modeler**: Create flows to prepare and blend data, build and manage models, and visualize the results.

- **Decision Optimization**: Evaluate all possible scenarios and identify the optimal solution to any given problem.

- **Data Refinery**: Simplify the process of preparing large amounts of raw data for analysis.

- **Cognos Dashboard Embedded**: Identify patterns in your data with sophisticated visualizations – no coding needed.

- **Analytics Engine for Apache Spark**: Automatically spin up lightweight, dedicated Apache Spark clusters to run a wide range of workloads.

- **Open Source Management**: Make it easy for developers and data scientists to find and access approved open source packages.

- **Netezza Performance Server**: A Netezza warehouse that is only available on defined deployment options (the Cloud Pak for Data system, AWS).

- **Match 360 with Watson**: A read-only copy of master data that enables customers to scale across the enterprise using RESTful APIs, flat files, or online. The AI-powered matching engine helps collapse data from multiple sources into a single source of truth.

- **Db2 Data Gate**: Extract, load, and synchronize mission-critical Db2 for z/OS data for high-volume transactional or analytic applications.

Premium IBM cartridges and third-party services

The following table shows the premium IBM services (also referred to as cartridges) that run on the Cloud Pak for Data platform but are priced and packaged separately. They primarily fall into two categories:

- Offerings with a significant install base:

    a. **Db2 AESE**: A relational database that delivers advanced data management and analytics capabilities for transactional workloads.

    b. **DataStage**: Deliver data at the right time to the right place with integration, transformation, and the delivery of data in batches and in real time.

    c. **Cognos Analytics**: Self-service analytics, infused with AI and machine learning, enabling you to create stunning visualizations and share your findings through dashboards and reports.

   d. **Information Server**: Get this highly scalable and flexible data integration solution that handles data volumes at scale.

   e. **Planning Analytics**: Easily create more accurate plans, budgets, and forecasts using data from across your business.

   f. **Master Data Management**: Empower business and IT users to collaborate and innovate with trusted master data. Establish a single view of data and enable users to deliver better business insights.

   g. **Open Pages**: Improve operational efficiency and drive down costs with an integrated governance risk and compliance solution leveraging Watson AI and advanced analytics.

- Born on the cloud offerings with no prior on-premises presence:

   a. **Watson Assistant**: Build conversational interfaces into any app, device, or channel.

   b. **Watson Discovery**: Find answers and uncover insights in your complex business content.

   c. **Watson API Kit**: Includes Watson Text to Speech, Speech to Text, Watson Knowledge Studio, Watson Language Translator, and so on.

   d. **Watson Financial Crimes Insights**: Simplify the process of detecting and mitigating financial crimes with AI and regulatory expertise.

   e. **Open Data for Industries**: Provides a reference implementation of a data platform supporting an industry-standard methodology for collecting and governing oil and gas data and serving that data to various applications and services. It helps integrate data silos and simplify data access to stakeholders.

Cloud Pak for Data also comes with a number of free accelerators and datasets that customers can use as a starting point. We will cover this in detail in the next section.

Industry accelerators

The industry accelerators that are provided by IBM are a set of artifacts that help you address common business issues. Each industry accelerator is designed to help you solve a specific business problem, whether it's preventing credit card fraud in the banking industry or optimizing the efficiency of your contact center.

All accelerators include a business glossary that consists of terms and categories for data governance. The terms and categories provide meaning to the accelerator and act as the information architecture for the accelerator.

Some accelerators also include a sample project with everything you need to analyze data, build a model, and display results. The sample projects include **detailed instructions**, **datasets**, **Jupyter notebooks**, **models**, and **R Shiny applications**. The sample projects can be leveraged as templates for your data science needs to learn specific techniques or to demonstrate the capabilities of AI and analytics services within Cloud Pak for Data.

The following are some industry accelerators that are included with Cloud Pak for Data:

- **Cross-Industry**:

  Contact Center Optimization: Improve the productivity of your customer contact center by describing and characterizing your contact center data.

  Customer 360 Degree View: Get a complete view of your customers by characterizing facets such as customer segmentation, credit risk, loyalty, and social media sentiment.

  Emergency Response Management: Optimize the routes and deployment of snowplows.

- **Banking and Financial**:

  Credit Card Fraud: Quickly detect credit card fraud to reduce financial losses and protect you and your customers.

  Customer Attrition Prediction: Discover why your customers are leaving.

  Customer Life Event Prediction: Plan ahead for the financial wellness of your client by reaching out with the right offer at the right time.

  Customer Offer Affinity: Identify the right financial products and investment opportunities for new and existing clients.

  Customer Segmentation: Easily differentiate between client segments by identifying patterns of behavior.

  Loan Default Analysis: Identify potential credit risks in your loan portfolio.

- **Energy and Utilities**:

  Demand Planning: Manage thermal systems to produce accurate energy volumes based on anticipated demand and energy generation.

  Utilities Customer Attrition Prediction: Discover why your customers are leaving.

  Utilities Customers Micro-Segmentation: Divide a company's customers into small groups based on their lifestyle and engagement behaviors.

  Utilities Demand Response Program Propensity: Identify which customers should be targeted for enrollment in the Demand response program.

  Utilities Payment Risk Prediction: Identify which customers are most likely to miss their payment this billing cycle.

- **Healthcare**:

  Healthcare Location Services Optimization: Determine how far patients will travel to access quality health care.

- **Insurance**:

  Insurance Claims: Process insurance claims more efficiently to save time and money.

- **Telco**:

  Intelligence Maintenance Prediction: Reduce your costs by scheduling maintenance at just the right time.

  Telco Churn: Predict a given customer's propensity to cancel their membership or subscription and recommend promotions and offers that might help retain the customer.

- **Manufacturing**:

  Manufacturing Analytics with Weather: Use machine learning models and The Weather Company dataset to help you understand the impact weather has on failure rate and identify actions that you can take to save time and money.

- **Retail**:

  Retail Predictive Analytics with Weather: Use machine learning models and The Weather Company data to help you understand how a retail inventory manager, marketer, and retail sales planner can quickly determine the optimal combination of store, product, and weather conditions to maximize revenue uplift, know what to keep in inventory, where to send a marketing offer, or provide a future financial outlook.

Sales Prediction Using The Weather Company Data: Use machine-learning models and The Weather Company data to help you predict how weather conditions impact business performance, for instance, prospective sales. The following table depicts all these in more detail:

Cross-industry	Contact Center Optimization: Improve the productivity of your customer contact center by describing and characterizing your contact center data.
Banking and Financial	Credit Card Fraud: Quickly detect credit card fraud to reduce financial losses and protect you and your customers.
Cross-Industry	360-Degree Customer View: Get a complete view of your customers by characterizing facets such as: Customer segmentation Credit risk Loyalty Social media sentiment
Banking and Financial	Customer Attrition Prediction: Discover why your customers are leaving.
Banking and Financial	Customer Life Event Prediction: Plan ahead for the financial wellness of your client by reaching out with the right offer at the right time.
Banking and Financial	Customer Offer Affinity: Identify the right financial products and investment opportunities for new and existing clients.
Banking and Financial	Customer Segmentation: Easily differentiate between client segments by identifying patterns of behavior.
Energy and Utilities	Demand Planning: Manage thermal systems to produce accurate energy volumes based on anticipated demand and energy generation.
Cross-Industry	Emergency Response Management: Optimize the routes and deployment of snowplows.
Healthcare	Healthcare Location Services Optimization: Determine how far patients will travel to access quality health care.
Insurance	Insurance Claims: Process insurance claims more efficiently to save time and money.
Telco	Intelligence Maintenance Prediction: Reduce your costs by scheduling maintenance at just the right time.
Banking and Financial	Loan Default Analysis: Identify potential credit risks in your loan portfolio.
Manufacturing	Manufacturing Analytics with Weather: Use machine learning models and The Weather Company data to help you understand the impact weather has on failure rate and identify actions that you can take to save time and money.

Retail	Retail Predictive Analytics with Weather: Use machine learning models and The Weather Company data to help you understand how a retail inventory manager, marketer, and retail sales planner can quickly determine the optimal combination of store, product, and weather conditions to maximize revenue uplift, know what to keep in inventory, where to send a marketing offer, or provide a future financial outlook.
Retail	Sales Prediction using The Weather Company Data: Use machine-learning models and The Weather Company data to help you predict how weather conditions impact business performance, for instance, prospective sales.
Telco	Telco Churn: Predict a given customer's propensity to cancel their membership or subscription and recommend promotions and offers that might help retain the customer.
Energy and Utilities	Utilities Customer Attrition Prediction: Discover why your customers are leaving.
Energy and Utilities	Utilities Customers Micro-Segmentation: Divide a company's customers into small groups based on their lifestyle and engagement behaviors.
Energy and Utilities	Utilities Demand Response Program Propensity: Identify which customers should be targeted for enrollment in the Demand Response Program.
Energy and Utilities	Utilities Payment Risk Prediction: Identify which customers are most likely to miss their payment this billing cycle.

Figure 2.9 – Sales Prediction using The Weather Company data

Let's now look into the packaging and deployment options of Cloud Pak for Data.

Packaging and deployment options

Cloud Pak for Data is available in three deployment options:

- Cloud Pak for Data software that can be deployed and managed on any public cloud or as a private cloud on-premises
- Cloud Pak for Data System, a hyper-converged system (includes compute, virtualization, storage, and networking in a single cluster) with optimized hardware and software
- Cloud Pak for Data "as a service" on IBM Cloud managed by IBM, providing a true SaaS experience

Cloud Pak for Data software has three editions:

- Enterprise edition
- Standard edition
- Non-production edition

While there are no functional differences between the three editions, the Standard edition has scalability restrictions while the non-production edition can only be used for development and QA deployments.

Red Hat OpenShift

OpenShift is a platform that allows enterprises to run containerized applications and workloads and is powered by Kubernetes under the covers. At its core, OpenShift is underpinned by open source software contributed by a wide community of developers and committers. As a result, there are different flavors of OpenShift available on the market. The open source project that actually powers OpenShift is called **Origin Kubernetes Distribution (OKD)** and you can start with that for free. OpenShift Container Platform, on the other hand, is Red Hat's commercial version and it comes with Red Hat support and enables your applications and workloads to be deployed on any public or private cloud.

What is Kubernetes?

Applications these days are increasingly built as discrete functional parts, each of which can be delivered as a container. This enables abstraction, agile development, and scalable deployments. However, it also means that for every application, there are more parts to manage. To handle this complexity at scale, we need a policy-driven orchestration platform that dictates how and where containers will run. Kubernetes (also known as K8s or Kube) is designed to handle exactly these challenges. It is the leading container orchestration framework that automates many of the manual processes involved in deploying, managing, and scaling containerized applications.

Kubernetes delivers significant benefits to developers, IT operations, and business owners and enables scalability, workload portability, and agile development. It was originally developed and designed by engineers at Google and is the secret sauce behind its Google Cloud services. Red Hat was one of the first companies to collaborate with Google on Kubernetes and continues to be a leading contributor. The Kubernetes project now belongs to the **Cloud-Native Computing Foundation (CNCF)**, owing to its donation by Google in 2015.

It's easy to think of **Origin Kubernetes Distribution (OKD)** as the upstream version of OpenShift and it is community-supported through mailing lists, GitHub, and Slack channels. It is totally free to use and modify as you see fit, unlike OpenShift which is supported by Red Hat staff and engineers. The following is a detailed list of differences between Kubernetes, OKD, and OpenShift Container Platform:

	Kubernetes	OKD	RedHat OpenShift
Platform			
Automated node configuration and cluster updates		X	X
Multi-host container scheduling	X	X	X
Self-Service provisioning	X	X	X
Service discovery	X	X	X
Enterprise operating system			X
Image Registry		X	X
Validated storage plugins		X	X
Validated networking plugins		X	X
Monitoring		X	X
Log aggregation		X	X
Multi-tenancy		X	X
Metering and chargeback			X
Developer Exterience			
Cloud service broker		X	X
Automated image builds		X	X
CI/CD and DevOps workflows		X	X
Validated third-party Kubernetes operators			X
Certified databases			X
Certified middleware			X
Certified ISV Solutions			X
Serverless applications (Knative)		X	X
Operations			
Built-on operational management			X
Zero downtime patching and upgrades			X
Enterprise 24/7 support			X
9-year support life cycle			X
Security response team			X

Figure 2.10 – Kubernetes versus OKD versus Red Hat OpenShift

While Kubernetes and OKD are good enough to get a development initiative started, enterprises looking to embrace and deploy containerized workloads in production benefit significantly by embracing OpenShift. Among other things, OpenShift helps with automation, enterprise security, management, partner ecosystems, and enterprise support – all critical ingredients for enterprise deployment. It comes packaged as part of Cloud Pak for Data and forms the infrastructure layer providing the common services needed. Also, Red Hat OpenShift enables the portability and multi-cloud support of Cloud Pak for Data, one of the platform's key differentiators.

Summary

In this chapter, you received a thorough introduction to Cloud Pak for Data, IBM's data and AI platform that helps enterprises implement "The AI Ladder" while leveraging modern cloud-native frameworks. In the next chapter, we will cover the "Collect" rung of the AI ladder in detail, focusing on evolving enterprise needs and how IBM is addressing them.

Section 2: Product Capabilities

In this section, we will learn about the product capabilities across the rungs of IBM's AI ladder and how Cloud Pak for Data helps deliver them as one integrated platform. We will review the IBM and third-party premium services available on Cloud Pak for Data, along with the most common use-case patterns of the platform and how Cloud Pak for Data addresses hybrid multi-cloud requirements.

This section comprises the following chapters:

3
Collect – Making Data Simple and Accessible

Enterprises are struggling with the proliferation of data in terms of both volume and variety. Data modernization addresses this key challenge and involves establishing a strong foundation of data by making it simple and accessible, regardless of where that data resides. Since data that's used in AI is often very dynamic and fluid with ever-expanding sources, virtualizing how data is collected is critical for clients. **Cloud Pak for Data** offers a flexible approach to address these modern challenges with a mix of proprietary, open source, and third-party services.

In this chapter, we will look at the importance of data and the challenges that occur with data-centric delivery. We will also look at what data virtualization is and how it can be used to simplify data access. Toward the end of this chapter, we will be looking at how Cloud Pak for Data enables **data estate modernization**.

In this chapter, we're going to cover the following main topics:

- Data – the world's most valuable asset
- Challenges with data-centric delivery
- Enterprise data architecture
- Data virtualization – accessing data anywhere
- *Data estate modernization* using Cloud Pak for Data

Data – the world's most valuable asset

The fact is, *every company in the world is a data company*. As the Economist magazine rightly pointed out in 2017, data (not oil) is the world's most valuable resource and unless you are leveraging your data as a strategic differentiator, you are likely missing out.

If you look at successful companies over time, all of them had sustainable competitive advantages – either economies of scale (Apple, Intel, AWS) or network effects (Facebook, Twitter, Uber, and so on). **Data** is the new basis for having a sustainable competitive advantage. Over 90% of the world's data cannot be googled, which means most of the world's valuable data is private to the organizations that own it. So, what can you do to unleash the potential that's inherent to your proprietary data?

As we discussed in *Chapter 1, Data Is the Fuel that Powers AI-Led Digital Transformation*, CEOs and business leaders know they need to harness digital transformation to jumpstart growth, speed up time to market, and foster innovation. In order to accelerate that transformation, they need to integrate processes across organizational boundaries by leveraging enterprise data as a strategic differentiator. It's critical to remember that your data is only accessible to you.

Data-centric enterprises

Let's look at a few examples of enterprises leveraging data as a strategic differentiator:

1. Best Buy successfully transformed its business by embracing digital marketing, along with personalized assistance and recommendations based on customer data. This digital transformation took multiple years, but ultimately, it helped the company survive and thrive as opposed to its competitor, Circuit City, which filed for bankruptcy and liquidated all its stores.

2. Uber is the world's largest taxi company, and the irony is that it doesn't own and operate a single taxi. The secret behind Uber's success is *data*. Uber employs its data as a strategic differentiator, driving all its business decisions, from prices to routes to determining driver bonuses. Consider Facebook, the world's most popular media owner that creates no content of its own; Alibaba, the world's most valuable retailer that has no inventory, and Airbnb, the world's largest accommodation provider with no real estate. The reason these companies are thriving and disrupting their competition is simple: they figured out a way to leverage data to drive their business decisions in real time.

3. Tik Tok's popularity and exponential growth over the past few years can be directly attributed to its recommendation engine, which leverages user data such as interactions, demographic information, and video content to drive highly accurate, personalized recommendations. Amazon and Netflix followed the same approach over the past decade to formulate personalized suggestions, which had a significant impact on their growth (and hence market capitalization).

Next, we will cover the challenges associated with data-centric delivery.

Challenges with data-centric delivery

Now that we have established that data is everywhere and that the best businesses in the world today are data-driven, let's look at what data-centric means. Enterprises are collecting data from more and increasingly diverse sources to analyze and drive their operations, with those sources perhaps numbering in the thousands or millions.

Here is an interesting fact: according to Forrester (`https://go.forrester.com/ blogs/hadoop-is-datas-darling-for-a-reason/`), up to 73% of the data you create in your enterprise goes unused. That's a very expensive and ineffective approach.

> **Note to Remember**
>
> Data under management is not the same as data stored. Data under management is data that can be consumed by the enterprise through a governed and common access point. This is something that hasn't really existed until today, but it is quickly becoming the leading indicator of a company's market capitalization.

And we are just getting started. Today, enterprises have roughly 800 terabytes of data under management. In 5 years, that number will explode to 5 petabytes. If things are complicated, slow, and expensive today, think about what will happen in 5 years.

The complexity, cost, time, and risk of error in collecting, governing, storing, processing, and analyzing that data centrally is also increasing exponentially. On the same note, the databases and repositories that are the sources of all of this data are more powerful, with abundant processing and data storage capabilities of their own available.

Historically, enterprises managed data through systems of record (mainframe and client/server applications). The number of applications was limited, and data that's generated was structured for the most part – relational databases were leveraged to persist the data. However, that paradigm is not valid anymore. With the explosion of mobile phones and social networks over the past decade, the number of applications and amount of data that's generated has increased exponentially. More importantly, the data is not structured anymore – relational databases, while still relevant for legacy applications, are not built to handle unstructured data produced in real time. This led to a new crop of data stores such as MongoDB, Postgres, CouchDB, GraphDB, and more, jointly referred to as NoSQL databases. These applications are categorized as **systems of engagement**:

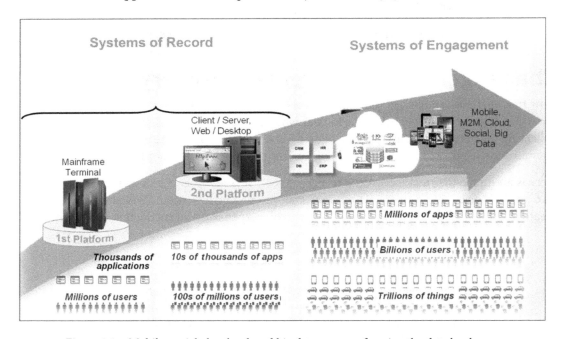

Figure 3.1 – Mobile, social, the cloud, and big data are transforming the data landscape

While the evolution of systems of engagement has led to an exponential increase in the volume, velocity, and variety of data, it has also made it equally challenging for enterprises to tap into their data, given it is now distributed across a wide number and variety of data stores within the organization:

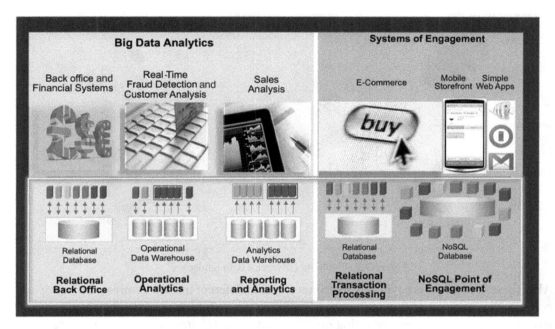

Figure 3.2 – Different workloads require different data stores

At the same time, business users are more sophisticated these days when it comes to creating and processing their own datasets, tapping into Excel and other desktop utilities to cater for their business requirements. Data warehouses and data lakes are unable to address the breadth and depth of the data landscape, and enterprises are beginning to look for more sophisticated solutions. To add to this challenge, we are beginning to see an explosion of smart devices in the market, which is bound to further disrupt the data architecture in industries such as manufacturing, distribution, healthcare, utilities, oil, gas, and many more. Enterprises must modernize and continue to reinvent to address these changing business conditions.

Before we explore potential solutions and how Cloud Pak for Data is addressing these challenges, let's review the typical enterprise data architecture and the inherent gaps that are yet to be addressed.

Enterprise data architecture

A typical enterprise today has several data stores, systems of record, data warehouses, data lakes, and end user applications, as depicted in the following diagram:

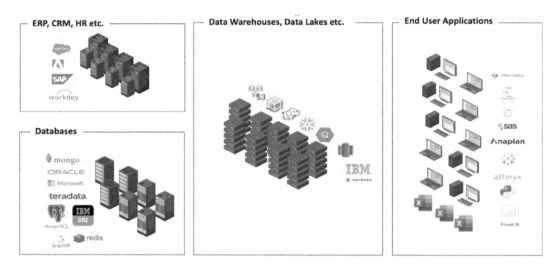

Figure 3.3 – Evolving enterprise data landscape

Also, these data stores are typically distributed across different infrastructures – a combination of on-premises and multiple public clouds. While most of the data is structured, increasingly, we are seeing unstructured and semi-structured datasets being persisted in NoSQL databases, Hadoop, or object stores. The evolving complexity and the various integration touchpoints are beginning to overwhelm enterprises, often making it a challenge for business users to find the right datasets for their business needs. This is represented in the following architecture diagram of a typical enterprise IT, wherein the data and its associated infrastructure is distributed, growing, and interconnected:

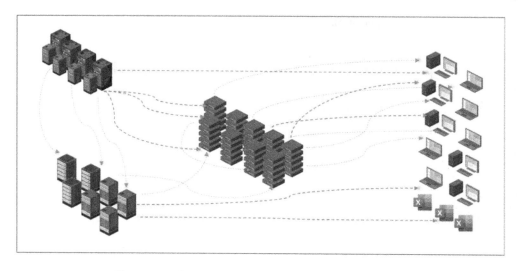

Figure 3.4 – Data-centric equates to increasing complexity

The following is a detailed list of all the popular data stores being embraced by enterprises, along with the specific requirements they address:

	Relational Databases	
1	Db2	Both transactional (OLTP) and analytical workloads.
2	Oracle Database	Famous among developers for being easy to use, having extensive documentation, and a rich feature set.
3	Microsoft SQL Server	#1 RDBMS in the world today. Used extensively as a basic data store for applications, as well as an enterprise-grade database.
4	IBM Informix	Known for its light footprint and time series data, which makes it a good match for IoT datasets.
5	MySQL (MariaDB)	Originally developed by Sun Microsystems and now part of Oracle, it includes a free community option and a supported commercial version. MariaDB is a commercial database built on open source MySQL.
6	Microsoft Access	A DBMS that combines the relational Microsoft Jet Database Engine with a graphical user interface.
7	SQLite	SQLite is a relational database management system contained in a C library.
8	Amazon RDS	A cost-effective cloud-based relational database that's compatible with six database engines, including PostgreSQL and Amazon Aurora.
9	PostgreSQL (Enterprise DB)	EnterpriseDB helps customers develop and run applications on a security-rich, enterprise-class database that's based on open source PostgreSQL.

Figure 3.5 – Relational databases

The following is a detailed list of NoSQL data stores.

NoSQL Data Stores		
8	MongoDB	A very popular document NoSQL database.
9	Redis	Popular key-value NoSQL database.
10	CouchDB	Document-based NoSQL database.
11	Cassandra	Popular wide-column NOSQL database.
12	MarkLogic	This schema-agnostic database lets you ingest data of any form or type, including geospatial data, JSON, RDF, and massive binaries such as videos. It has a built-in search engine that's continuously being optimized for queries.
13	Elasticsearch	Elasticsearch is a highly scalable, open source, full-text search and analytics engine for storing, searching, and analyzing big volumes of data in near real time.
14	Neo4j	Well-known graph NoSQL database.
15	Memcached	Key-value NoSQL database.
16	Riak DB	Key-value NoSQL database.
17	Raven DB	Document-based NoSQL database.
18	Apache HBase	Wide-column NoSQL database.

Figure 3.6 – NoSQL Datastores

The following is a detailed list of data warehouses and data lakes:

Data Warehouses and Data Lakes		
19	IBM Netezza	This is a very popular data warehouse appliance that is known for its ease of use, scalability, and performance.
20	Cloudera Data Platform (CDP)	CPD represents a major step forward toward combining the value-added distributions of Hadoop from both Cloudera (CDH) and Hortonworks (HDP) into a unified, cloud-ready data and analytics platform.
21	Teradata	A data warehousing platform for collecting and analyzing vast amounts of enterprise data, all while employing smart in-memory processing to optimize database performance.
22	Oracle Exadata	Exadata is a pre-configured combination of hardware and software that provides an infrastructure solely for running Oracle Database.
23	Snowflake	Snowflake's multi-tenant, shared architecture separates storage from processing power, allowing customers to scale resources based on usage and performance requirements.
24	Amazon Redshift	Amazon Redshift is a data warehouse product that forms part of Amazon Web Services (AWS). The name means to shift away from Oracle, with red being an allusion to Oracle.
25	Google BigQuery	BigQuery is Google's data warehouse offering with built-in integrations for ML and TensorFlow, allowing you to quickly create powerful AI models. It can scale to petabytes of data for real-time analytics and supports geospatial analytics. Its supports separation of compute and storage, enabling you to scale processing and memory resources separately.
26	Amazon S3	Addresses cloud storage needs cheaply at scale for big data analytics – it's an object-oriented service and stores data in "buckets," each of which can hold up to 5 terabytes.
27	IBM Db2 Warehouse	Suited for analytics and AI workloads, Db2 Warehouse provides built-in machine learning tools and supports SQL and Python languages with an intuitive UI and REST APIs. Its MPP capabilities allow for fast, concurrent querying across large datasets.

Figure 3.7 – Data warehouses and data lakes

Now that we have covered enterprise data architecture, let's move on to NoSQL data stores.

NoSQL data stores – key categories

NoSQL data stores can be broadly classified into four categories based on their architecture and usage. They are as follows:

- **Document databases** map each key with a complex data structure, called a document. Documents can be key-array pairs, key-value pairs, or even nested documents. Some of the most popular choices for document databases include MongoDB, CouchDB, Raven DB, IBM Domino, and MarkLogic.

- **Key-value stores** are the *simplest* form of NoSQL database. Every row is stored as a key-value pair. Popular examples include Redis, Memcached, and Riak.

- **Wide-column stores**, also known as columnar data stores, are optimized for making queries over large datasets, as well as to store columns of data together instead of as rows. Examples include Cassandra and HBase.

- **Graph stores** store information about graphs; networks, such as social connections; road maps; and transport links. Neo4j is the most popular graph store.

Next, we'll review some of the capabilities of Cloud Pak for Data that enable it to address some of the challenges and one of them is Data virtualization.

Data virtualization – accessing data anywhere

Historically, enterprises have consolidated data from multiple sources into central data stores, such as *data marts*, *data warehouses*, and *data lakes*, for analysis. While this is still very relevant for certain use cases, the time, money, and resources required make it prohibitive to scale every time a business user or data scientist needs new data. Extracting, transforming, and consolidating data is resource-intensive, expensive, and time-consuming and can be avoided through data virtualization.

Data virtualization enables users to tap into data at the source, removing complexity and the manual processes of data governance and security, as well as incremental storage requirements. This also helps simplify application development and infuses agility. **Extract, Transform, and Load** (ETL), on the other hand, is helpful for complex transformational processes and nicely complements data virtualization, which allows users to bypass many of the early rounds of data movement and transformation, thus providing an integrated, business-friendly view in near real time.

The following diagram shows how data virtualization connects data sources and data consumers, thereby enabling a single pane of glass to access distributed datasets across the enterprise:

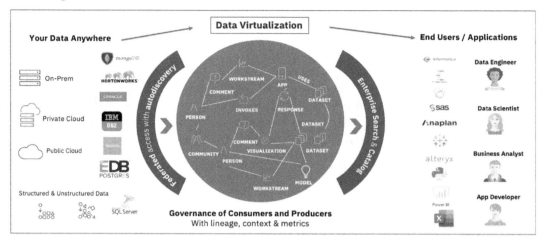

Figure 3.8 – Data virtualization – managing all your data, regardless of where it resides

There are very few vendors that offer data virtualization. According to Tech Target, the top five vendors in the market for data virtualization are as follows:

- Actifio Sky
- Denodo
- IBM Cloud Pak for Data
- Informatica PowerCenter
- Tibco Data Virtualization

Among these, IBM stands out for its integrated approach and scale. Its focus on a unified experience of bringing data management, data governance, and data analysis into a single platform resonates with today's enterprise needs, while its unique IP enables IBM's data virtualization to scale both horizontally and vertically. Among other things, IBM leverages push-down optimization to tap into the resources of the data sources, enabling it to scale without constraints.

Data virtualization connects all the data sources to a single, self-balancing collection of data sources or databases, referred to as a *constellation*. No longer are analytics queries performed on data that's been copied and stored in a centralized location. The analytics application submits a query that's processed on the server where the data source exists. The results of the query are consolidated within the constellation and returned to the original application. No data is copied, and it only exists at the source.

By using the processing power of every data source and accessing the data that each data source has physically stored, latency from moving and copying data is avoided. In addition, all repository data is accessible in real time, and governance and erroneous data issues are virtually eliminated. There's no need for extract, transform, and load and duplicate data storage, accelerating processing times. This process brings real-time insights to decision-making applications or analysts more quickly and dependably than existing methods. It also remains highly complementary with existing methods and can easily coexist when it remains necessary to copy and move some data for historical, archival, or regulatory purposes:

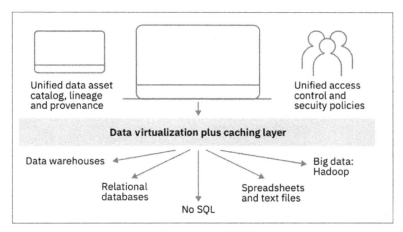

Figure 3.9 – Data virtualization in Cloud Pak for Data

A common scenario in distributed data systems is that many databases store data in a common schema. For example, you may have multiple databases storing sales data or transactional data, each for a set of tenants or a region. Data virtualization in Cloud Pak for Data can automatically detect common schemas across systems and allow them to appear as a single schema in data virtualization – a process known as **schema folding**. For example, a SALES table that exists in each of the 20 databases can now appear as a single SALES table and can be queried through **Structured Query Language** (**SQL**) as one virtual table.

Data virtualization versus ETL – when to use what?

Historically, Data warehouses and data lakes are built by moving data in bulk using ETL. One of the leading ETL products in the market happens to be from IBM and is called IBM DataStage. So, it begs the question as to when someone should use data virtualization versus an ETL offering. The answer depends on the use case. If the intent is to explore and analyze small sets of data in real time and where data can change every few minutes or hours, data virtualization is recommended. Please note that the reference to small sets of data alludes to the actual data that's transferred, not the dataset that a query is performed on. On the flip side, if the use case requires processing huge datasets across multiple sources and where data is more or less static over time (historical datasets), an ETL-based solution is highly recommended.

Platform connections – streamlining data connectivity

Cloud Pak for Data also includes the concept of **platform connections**, which enable enterprises to define data source connections universally. These can then be shared by all the different services in the platform. This not only simplifies administration but enables users across different services to easily find and connect to the data sources. Cloud Pak for Data also offers flexibility to define connections at the service level to override platform connections. Finally, administrators can opt to either set user credentials for a given connection or force individual users to enter their respective credentials. The following is a screenshot of the platform connections capability of **Cloud Pak for Data v4.0**, which was released on June 23, 2021:

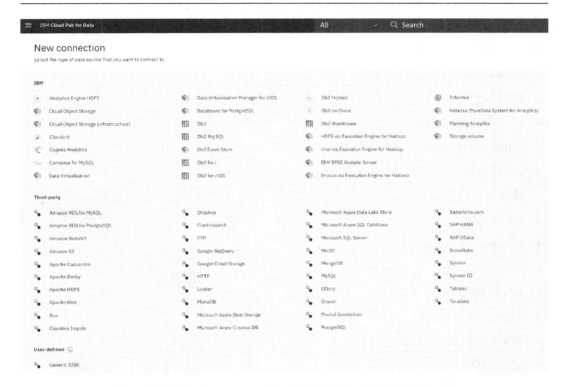

Figure 3.10 – Platform connections in IBM Cloud Pak for Data v4.0

Now that we have gone through the collect capabilities of Cloud Pak for Data, let's see how customers can modernize their data estates for Cloud Pak for Data and the underlying benefits of this.

Data estate modernization using Cloud Pak for Data

So far, we have seen the evolving complexity of the data landscape and the challenges enterprises are trying to address. Increasing data volumes, expanding data stores, and hybrid multi-cloud deployment scenarios have made it very challenging to consolidate data for analysis. IBM's Cloud Pak for Data offers a very modern solution to this challenge. At its core is the data virtualization service, which lets customers tap into the data in source systems without moving the data. More importantly, its integration with the enterprise catalog means that any data that's accessed is automatically discovered, profiled, and cataloged for future searches. Customers can join data from multiple sources into virtualized views and can easily enforce governance and privacy policies, making it a one-stop shop for data access. Finally, its ability to scale and leverage source system resources is extremely powerful.

IBM's data virtualization service in Cloud Pak for Data supports over 90% of the enterprise data landscape:

- Apache Hive
- IBM Big SQL
- Cloudera Impala
- Third-party custom JDBC drivers
- Db2 Event Store
- Db2 Warehouse
- Db2 Database
- Db2 for z/OS
- Hive JDBC
- Informix
- MariaDB
- Microsoft SQL Server
- MongoDB
- MySQL Community Edition
- MySQL Enterprise Edition
- Netezza
- Oracle
- PostgreSQL
- Teradata (requires Teradata JDBC drivers to connect)
- Sybase IQ
- Amazon Redshift
- Google BigQuery
- Pivotal Greenplum
- Salesforce.com
- SAP HANA and SAP ODATA
- Snowflake

While tapping into data without moving it is a great starting point, Cloud Pak for Data also enables customers to persist data on its platform. You can easily deploy, provision, and scale a variety of data stores on Cloud Pak for Data. The supported databases on Cloud Pak for Data are as follows:

- Db2 (both OLTP and analytical workloads)
- Db2 Event Store
- MongoDB
- Enterprise DB PostgreSQL
- Cockroach DB
- Crunchy DB PostgreSQL
- Support for object storage – separation of compute and storage

Cloud Pak for Data enjoys a vibrant and open ecosystem, and more data sources are scheduled to be onboarded over the next 1-2 years. This offers customers the freedom to embrace the data stores of their choice while continuing to tap into their existing data landscape. Finally, it's worth mentioning that all the data stores that are available on the platform are containerized and cloud-native by design, allowing customers to easily provision, upgrade, scale, and manage their data stores. Also, customers can deploy these data stores on any private or public cloud, which enables portability. This is critical in situations where data gravity and the co-location of data and analytics is critical.

Summary

Data is the world's most valuable resource and to be successful, enterprises need to become data-centric and leverage their data as a key differentiator. However, increasing data volumes, evolving data stores, and distributed datasets are making it difficult for enterprises and business users to easily find and access the data they need. Today's enterprise data architecture is fairly complex, with a plethora of data stores optimized for specific workloads. Data virtualization offers the silver bullet to address this unique challenge, and IBM's Cloud Pak Data is one of the key vendors in the market today. It is differentiated for its ability to scale and its integrated approach, which addresses data management, data organization, and data analysis requirements.

Finally, IBM's Cloud Pak for Data complements its data virtualization service with several containerized data stores that allow customers to persist data on the platform, while also allowing them to access their existing data without moving it.

In the next chapter, you will learn how to create a trusted analytics foundation and organize the data you collected across your data stores.

4

Organize – Creating a Trusted Analytics Foundation

Most enterprises are going through the digital transformation journey to better understand their customers and deliver products and services that suit their needs. This has led to a proliferation of applications/systems that aim to improve the customer experience. Consequently, a large, diverse dataset has been made available by these applications/systems.

This presents us with a unique opportunity to leverage data for better decision-making, but this also poses a challenge in finding the data when we need it. Furthermore, how do we decide what dataset would best fit our needs? Are we able to trust the quality of the data? These questions are becoming the norm, and enterprises are grappling to address them as they focus on tapping into their most valuable resource: data. The answer is the evolution of a new practice called **DataOps**.

Cloud Pak for Data's capabilities help with establishing a governance framework around the information assets in the enterprise. This helps with being able to find data but more importantly, it helps enrich the information about the data we have in a centralized governance catalog, which enables access to trusted data.

In this chapter, we're going to cover the following main topics:

- Introducing **Data Operations (DataOps)**
- Organizing enterprise information assets
- Establishing metadata and stewardship
- Profiling to get a better understanding of your data
- Classifying data for completeness
- Enabling trust with data quality
- Data privacy and activity monitoring
- Data integration at scale
- IBM Master Data Management
- Extending Master Data Management toward a digital twin

Introducing Data Operations (DataOps)

Data is the fuel for innovation and sustaining a competitive advantage. It is the key ingredient for driving analytics and understanding business trends and opportunities. Unlocking this value in new ways can accelerate an organization's journey to AI.

DataOps is focused on enabling collaboration across an organization to drive agility, speed, and new data initiatives at scale. By using the power of automation, DataOps is designed to solve challenges associated with inefficiencies in accessing, preparing, integrating, and making data available.

At the core of DataOps is an organization's information architecture, which addresses the following questions:

- Do you know your data? Do you trust your data?
- Are you able to quickly detect errors?
- Can you make changes incrementally without "breaking" your entire data pipeline?

To answer all these questions, the first step is to take inventory of your data governance practices. The following diagram illustrates the automation that's introduced by DataOps in the data curation process:

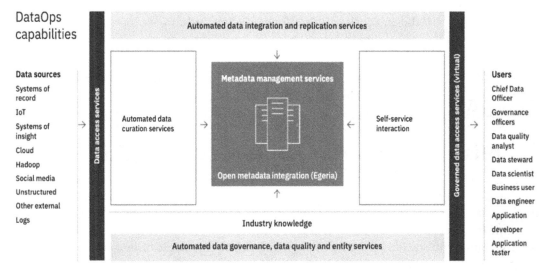

Figure 4.1 – DataOps toolchain

While establishing the DataOps practice, the five key considerations are as follows:

- **Data curation**
- **Metadata management**
- **Data governance**
- **Master data management**
- **Self-service interaction**

The first four of these considerations will be covered in this chapter. We delve into the self-service interaction as part of *Chapter 5, Analyze: Build, Deploy, and Scale Models with Trust and Transparency*.

Organizing enterprise information assets

Enterprise information assets are the center of the universe from which organizations derive their knowledge to conduct their business efficiently, in compliance with any/all regulatory requirements they are bound to, and to serve their customers more effectively.

Information assets can be broadly classified as *business assets* and *technical assets*. **Business assets** include the taxonomy of business terminology, organizational governance policies, and the governance rules that help enforce governance policies, while **technical assets** include all assets that relate to enterprise applications, the underlying data repositories, data files that support the enterprise applications, and any middleware technologies that enable independent applications and services to exchange information.

It is important to bring these two pieces of metadata together to have an effective organization that allows the business audience to communicate easily with the technical audience, who will be responsible for the underlying applications. In the following sections, we will walk through the considerations for organizing enterprise information assets.

Establishing metadata and stewardship

We discussed information assets in the previous section; that is, identifying business and technical assets as two components that make up the information assets. **Metadata** is data about data. In our case, metadata is information that relates to the business assets and technical assets in an enterprise. Think of the last time you interacted with a bot or a search engine: you started by typing in keywords about your topic of interest and then thought of additional criteria to narrow down your search. Or, the search engine gave you faceted search options to filter out the noise from the information. The keywords you used – both the ones you started with and the recommendations based on faceted search – are metadata that describes the desired outcome.

Capturing metadata is an important and first step toward establishing a governance framework. It is important because it enhances the understanding we have about the enterprise information assets and enables consumers of the data to search and find the assets that meet their needs. Complementary information is considered metadata when it helps answer the following about the information asset:

- What
- When
- Where
- Who
- How
- Which
- Why

Data governance is about defining and managing organizational policies, procedures, and standards both to ensure consistency and, more importantly, to cater to the regulatory requirements. Metadata is crucial for data governance, but another equally important concept for data governance is stewardship.

Stewardship is about aligning information assets with personnel across different departments and business units. Stewards are entrusted with managing one or more enterprise assets and ensuring the currency and quality of the asset. The core responsibilities of a data steward include the following:

- Identifying subject areas and logical areas to organize metadata
- Defining taxonomy of business categories and terms
- Defining reference data
- Establishing governance policies and rules
- Ensuring data quality
- Establishing data security policies
- Acting as a conduit between the business and technology
- Helping identify data integration issues:

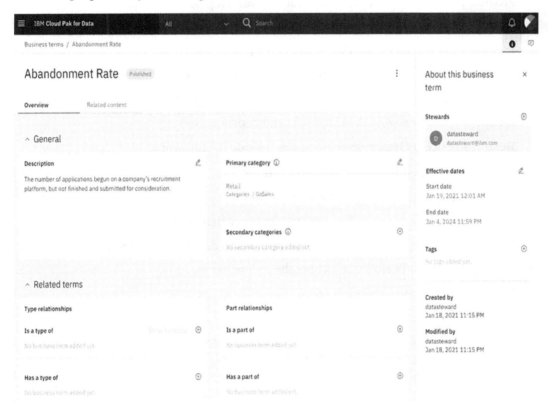

Figure 4.2 – Stewardship of business metadata

The previous screenshot shows an example of business metadata that's been collected in the form of business terms, with a data steward assigned. They will be responsible for validating the business term and organizing this business concept in the enterprise taxonomy through categories and secondary categories where applicable.

Data stewards play a key role in the broader data governance initiative. Let's use a simple diagram to illustrate the roles that are played by the different entities, including their individual responsibilities:

- **Data Governance Leadership** responsibilities include the following:

    Formulating and reviewing governance plans and policies

    Being aware of leading trends in data governance areas

    Providing guiding principles for various data governance initiatives

    Providing strategic direction for the data stewardship committee

- **Data Governance Office** responsibilities include the following:

    Identifying and prioritizing data governance initiatives

    Establishing and monitoring performance measures for data

    Providing assurance of the enterprise perspective and approach

    Identifying and engaging data stewards and coordinators

- **Business Process Owner** responsibilities include the following:

    Resolving escalated problems at a tactical level

    Identifying and engaging operational data stewards

    Prioritizing and approving data quality issue remediation requests

    Communicating expectations and requirements to data stewards and custodians

- **Data Steward** responsibilities include the following:

    Using data to perform their job and processes.

    Checking the integrity of data usage.

    Serving as the first point of contact for data access, source, change control, and any data quality issues that are identified within their data domain and functional area.

Recognizing triggers for data governance activities:

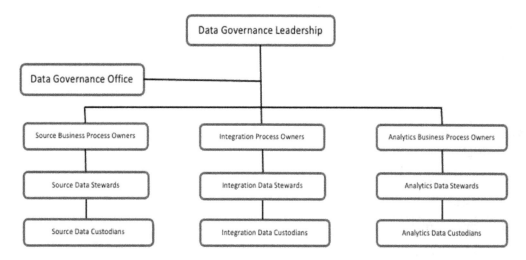

Figure 4.3 – Data governance organization

Now that we understand how the governance team is organized, we will enlist the most commonly used business and technical metadata components.

Business metadata components

The following are the most commonly used business and technical metadata components:

- Category hierarchy
- Categories
- Business terms
- Business term type hierarchy
- Business term types
- Information governance policy hierarchy
- Information governance policies
- Information governance rules

Now, we will move on and cover the technical metadata components.

Technical metadata components

Technical metadata can be both out-of-the-box asset types and also custom asset types. This includes the following:

- Databases
- Data files
- Unstructured data sources
- Data science assets (models, notebooks, and so on)
- Logical data models
- Physical data models
- XML schema definitions
- Master data management
- Applications
- Files
- Stored procedure definitions
- Business intelligence artifacts
- Data quality assets (data rule definitions, data rule sets, metrics, and so on)
- Extraction, Transformation, Loading assets (jobs, stages, parameters, and so on)
- Custom technical assets

Now that we know what metadata needs to be captured, in the next section, we will discuss the process of getting a better understanding of the data we have by profiling the enterprise data assets.

Profiling to get a better understanding of your data

With enterprises going through digital transformation, there has been a proliferation of data and multiple systems that often result in redundant datasets with varying levels of data quality. With large volumes of data, it is imperative that we get a good understanding of the characteristics of the datasets, what attributes/columns of data are available, the data types of the individual columns, their unique values, and how the data values are distributed. This profile information helps us isolate datasets of interest from the large collection of possibly overlapping datasets.

With Cloud Pak for Data, we have a centralized governance catalog that acts as the index of all our data assets and helps us organize resources for many data science projects: data assets, analytical assets, and the users who need to use these assets. This catalog also has built-in profiling capabilities that allow the consumers to select a given dataset and initiate a quick data profiling activity. The focus is on the following two aspects of profiling:

- Data structure discovery; that is, what attributes are available in the dataset?
- Data content discovery; that is, what are the data values, their distribution, quality, and so on?

The first step is to define a connection to the data source and onboard the data asset to the catalog. Cloud Pak for Data supports a wide array of data source connectors, and this list of supported connectors is growing with every new version of the platform.

Once a data asset has been onboarded, we can use the catalog to do the following:

- Tag the asset with one or more tags.
- Get a quick preview of the data.
- Add reviews and comment on the asset.
- Define access policies for the asset.
- Track the event lineage of the asset.
- Create a data profile:

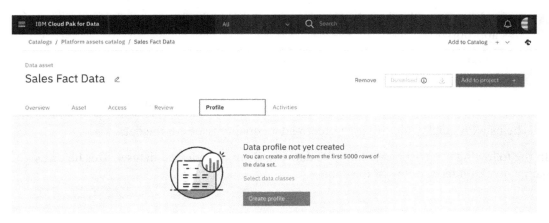

Figure 4.4 – Initial profiling of catalog assets

Optionally, you can select either all the available data classes or a subset of the data classes as part of the profiling activity. The amount of time it takes to run the profile will depend on the number of data classes we pick and the size and width of the data asset:

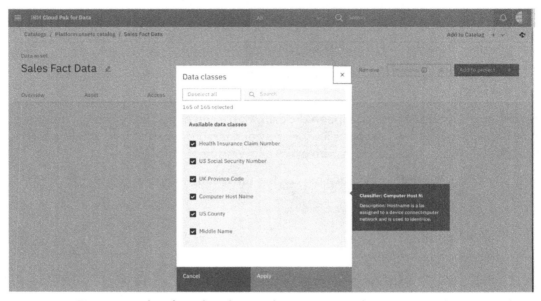

Figure 4.5 – Identifying data classes to leverage as part of initial data profiling

The data profiling process takes some time and you get to see a few things:

- The data classes that have been assigned to individual columns/attributes as appropriate

- The summary statistics of the individual columns/attributes, regardless of the data type

- The frequency distribution of the values in the individual columns/attributes

This helps with quickly identifying potential data quality issues in the data.

In the following screenshot, we have the data profile of the selected dataset. This helps us gain a quick understanding of the data:

Figure 4.6 – Initial profiling results showing data distribution and summary statistics

We also have a more comprehensive profiling capability as part of the auto-discovery and data quality analysis functions, which we will discuss later in this chapter. This will take us further in terms of discovery relationships across data domains and doing a more comprehensive data quality analysis.

Classifying data for completeness

Classifications are special labels that are used to classify assets based on the enterprise confidentiality level. This is done by tagging certain assets as being more sensitive than others. You can think of them as tags that help consumers identify what assets they can share, as well as protect other assets that are not available for distribution outside the business unit or organization. The centralized catalog service in Cloud Pak for Data comes with the following out-of-the-box classifications, with the flexibility to add more classifications as required by the organization:

- **Personally Identifiable Information** (**PII**), which is data that can identify individuals (for example, social security numbers).

- **Sensitive Personal Information** (**SPI**), which is data that, if exposed, can cause substantial harm, embarrassment, and/or unfairness to an individual (for example, race, sexual orientation, religious beliefs, and so on).

- **Confidential**, which is data that, if compromised, can lead to long-term harm to the individual and/or organization:

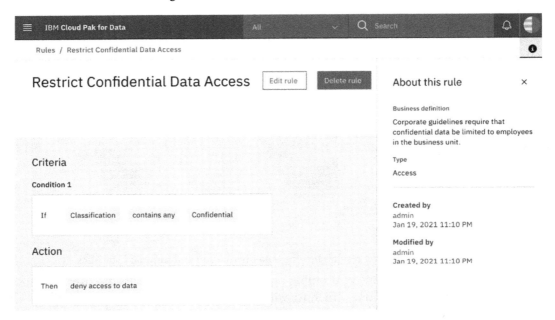

Figure 4.7 – Classification used to define data protection rules

Not only can we use these classifications as asset tags, but we can also leverage the platform's automation to trigger data protection rules that would restrict access to certain assets for different users based on their classification.

Next, we will discuss how we can leverage the power of Cloud Pak for Data's automated data discovery feature and its **machine learning** (**ML**) infused business term assignment process.

Automating data discovery and business term assignment

So far, we have discussed different types of metadata (business and technical) and the value of metadata in organizations. One crucial task is to map business metadata to technical metadata. This is required because Cloud Pak for Data is a platform that caters to different personas, including business and technical audiences, and the goal is to facilitate communication between the business and technical audiences, in addition to aligning technical assets to the corresponding business functions.

Cloud Pak for Data has a patented ML-based auto-classification function that recommends mapping business terms to the technical assets as part of the auto-discovery process. The ML classification model is trained with industry-specific metadata and provides cognitive learning based on manual business term assignments that are done by subject matter experts.

Automatic term assignment is invoked as part of column analysis in the discovery process. We have two types of discovery processes:

- QuickScan
- Auto-discovery

Terms are automatically assigned based on the confidence level of the ML model's scoring. These associations of the business terms to technical assets are initially represented as candidates that domain experts and stewards can review and assign manually. When the confidence level matches or exceeds 50%, terms are suggested to be assigned. When the confidence level exceeds 80%, candidates are automatically assigned.

The ML model uses the following to generate term assignments:

- Linguistic name matching, based on the similarity of the name of the business term and the data asset.
- Data class-based assignment. In this case, if a business term is associated with a data class and if data asset profiling assigns the data class to an asset, then the term is recommended as a match to the asset.
- A supervised learning model, which is initially trained with terms stored in the default catalog and subsequently learns from user term assignments as model feedback.

We will showcase the quickscan discovery process to showcase this automatic term assignment. We will invoke a quickscan discovery job on a data source connection that has already been defined on the platform. As part of this job, we can specify the following sub-tasks:

1. Perform column analysis (analyze columns).
2. Perform data quality analysis (analyze data quality).
3. Perform automatic term assignment (assign terms).

4. Use data sampling:

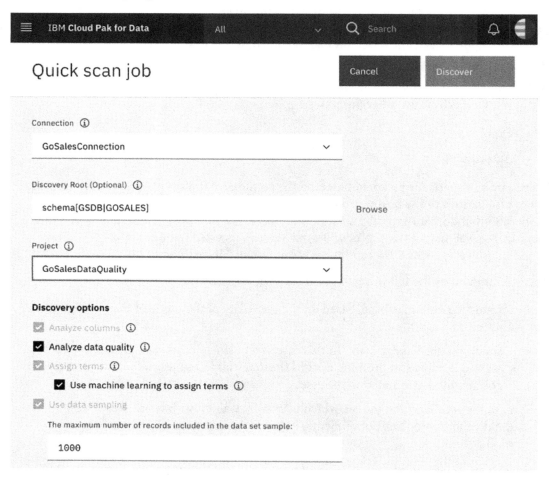

Figure 4.8 – Invoking the quick scan discovery job

The execution duration of **Quick scan job** can vary, depending on the number of data source tables and the number of data samples we use. In our case, it took about 50 seconds to discover 30 tables, which is very quick. In the following screenshot, you can see a summary of what automatic business term assignment it was able to accomplish, as well as the data class assignments for the individual columns/attributes:

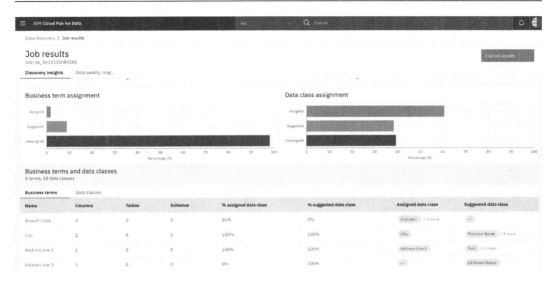

Figure 4.9 – Quick scan summary showing automatic term assignments for multiple columns

On the details screen, we can see the individual data asset columns, the derived data quality, the data class assignment and the suggested business terms, and the assigned business terms based on the confidence of the ML model scoring results. There is also a way for the subject matter expert to manually alter the term assignment suggestion. Any changes that are made here by the steward will result in term assignment model retraining. So, the next time we run the discovery job, it will recommend a different term assignment:

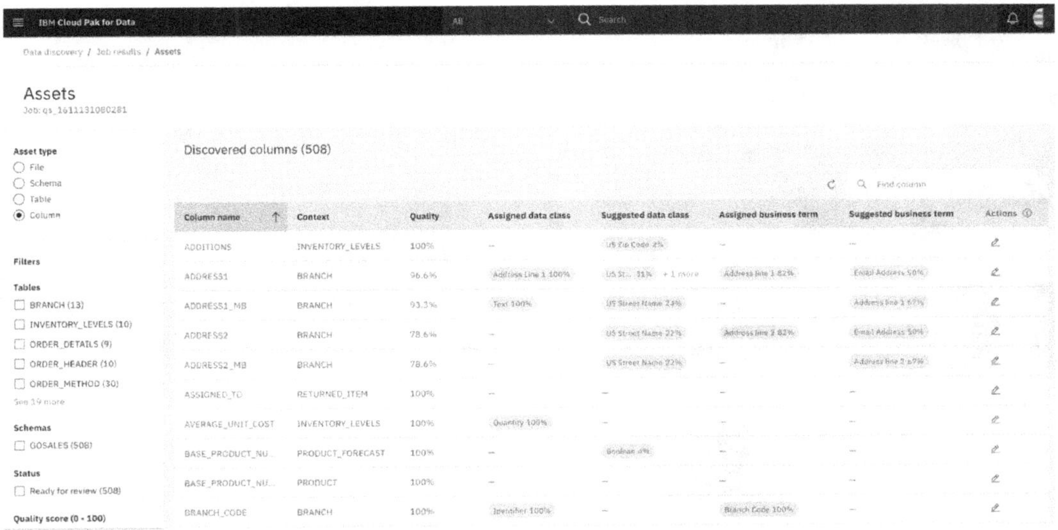

Figure 4.10 – Quick scan detail report of automatic term assignments

Upon reviewing the results of the discovery process, we now have the opportunity to protect sensitive data assets using the **IBM Guardium** integration in Cloud Pak for Data:

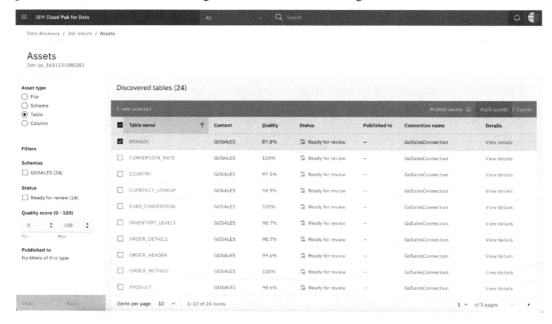

Figure 4.11 – Audit assets resulting from the asset discovery process

The following diagram illustrates the high-level process that should be followed to audit assets:

Figure 4.12 – High-level process for handling sensitive data elements

We have now seen how we can leverage automation around data discovery and how to assign business terms to technical assets. Next, we will discuss the data quality capabilities of the platform.

Enabling trust with data quality

Data quality is a cornerstone of business operations in the organization, and it is affected by the way data is entered, stored, and managed. The reason for a lack of confidence in information is because information is pervasive across the organization. We are dealing with fragmented silos of data that were accumulated through many years, without data quality measures and without being organized in a way that makes sense to the business.

High-quality data is essential for high-quality outcomes for any analytics we perform using the data. Data quality entails the following:

- **Completeness**: Having a good understanding of all the related data assets.
- **Accuracy**: Common data problems such as missing values, incorrect reference data, and so on that must be eliminated so that we have consistent data.
- **Availability**: Data must be available on demand.
- **Timeliness**: The up-to-dateness of the data is crucial to making the right decisions.

The data quality analysis area of Cloud Pak for Data performs the following tasks:

- Runs column analysis
- Evaluates quality dimensions to identify data quality problems
- Computes a data quality score for data assets and columns
- Runs quality rules and automation rules
- Optionally captures sample data with data quality problems

Upon competing the data quality analysis, you can publish the results to the governance catalog so that the other consumers of the platform can benefit from this analysis.

Steps to assess data quality

Let's walk through the data quality assessment capabilities by using the auto-discovery capabilities of the platform. The high-level steps we will go through for one of the possible scenarios are as follows:

1. Define a connection to the data source.

2. Initiate the auto-discovery process by selecting one or more of the following options:

    Analyze columns.

    Analyze data quality.

    Assign terms.

    Publish results to the catalog.

3. Once the data assets have been discovered, we must review the following:

    Profiling results

    Data quality analysis

    Data classifications

    Term assignments

4. We will then look at the data quality project and look at the details of the data quality analysis process:

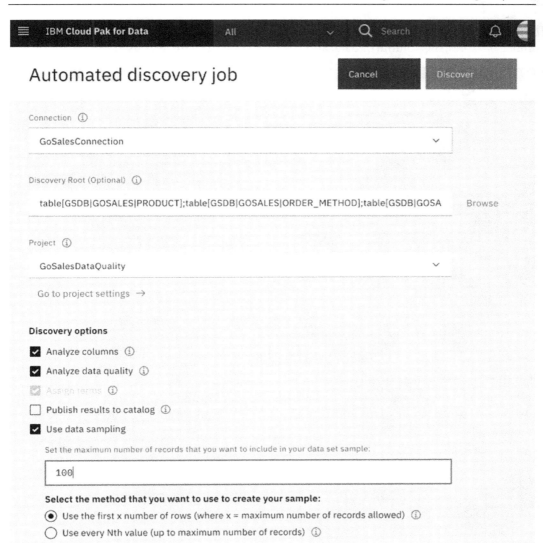

Figure 4.13 – Initiating an automated discovery job

The amount of time it takes for the auto-discovery job to complete depends on the number of tables, the volume of data, and whether data sampling was used:

Figure 4.14 – Data quality dashboard

Attaining the desired data quality is an evolving process. Here, we learn over time and establish a collection of data quality rules to handle data conditions, monitor data quality metrics, and fix our applications and/or source systems until we achieve the desired data quality results:

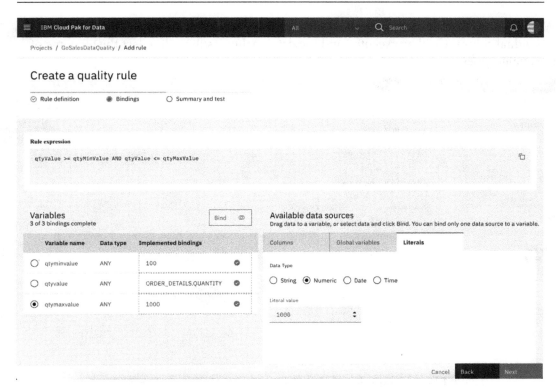

Figure 4.15 – Quality rule to identity any order quantity <100 or >1,000

When an auto-discover job is triggered, all the quality rules that are associated with the individual columns will be executed and the data quality results will reflect the outcome of the quality rules:

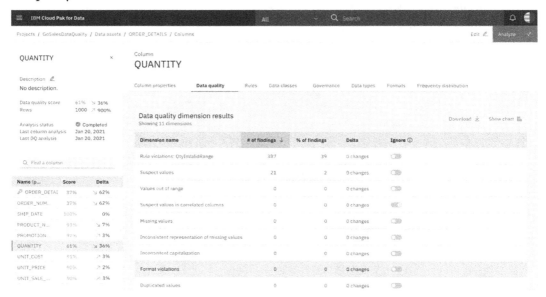

Figure 4.16 – Quality rule violation captured as part of the discovery process

Data quality is important to ensure our outcomes/analytics are of the highest possible quality. However, it does not stop there. You need to have a good understanding of the types of data assets and how data is distributed to quickly assess which datasets to use for analytics. In the next section, we will learn how to leverage DataOps for this purpose.

DataOps in action

We now have a good, foundational understanding of the different aspects of governance, metadata, data, and the operations we can perform on data, including data profiling and quality assessment. Now, let's look at a sample flow of information and the interactions across the different personas:

1. This process always starts with collecting raw data from the different source systems.

2. From the staging area, data analysts do some initial profiling and cleansing of data.

3. This is followed by more comprehensive data transformation before the data is loaded into an enterprise data warehouse and possibly some data marts.

4. Finally, we must get to a business-ready state where data is ready to be consumed by the end users for analytics:

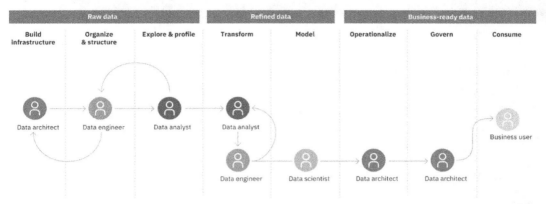

Figure 4.17 – Example DataOps workflow by personas

We now understand the overall process the data goes through and the interactions that occur across the different personas. Now, let's explore how we can bring some automation to data quality assessment.

Automation rules around data quality

Data quality assessment evaluates data against a set of criteria. This set of criteria depends on the business context of the data, including what information the data represents, what the criticality of the information is, and the purpose of the information. Data quality dimensions and custom data quality rules should be used together to perform data quality assessments. Data profiling provides an initial view of the characteristics and distribution of data. Automation rules leverage this information, from data profiling to assessing quality, such as checking for Null or valid date formats, and so on.

Cloud Pak for Data allows us to construct and deploy automation rules. These automation rules help automate tasks to ensure we have the highest quality data and include the following:

- Applying rule definitions based on data quality dimensions
- Assigning terms or stewards to assets

- Analyzing assets, and many more:

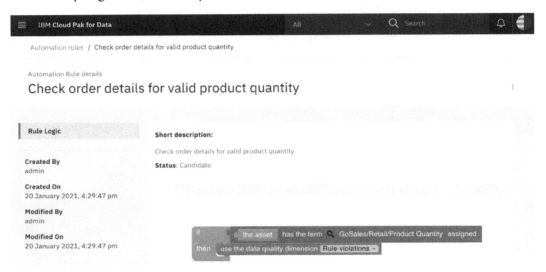

Figure 4.18 – Automation rule to conditionally apply quality rules

Now that we've learned how to improve data quality with automation, let's learn how data security needs are addressed by the data privacy and data activity monitoring capabilities of Cloud Pak for Data.

Data privacy and activity monitoring

Increasing news headlines about data breaches and/or theft of personal information and individual identities have brought the focus back on protecting data assets. As a consequence of this, we also now have several regulations being enacted around the world, such as **California Consumer Privacy Act (CCPA)**, **General Data Protection Regulation (GDPR)**, and more. Although the specifics of each of these regulations might vary, the central theme is that failing to protect data and ensuring data privacy could result in expensive consequences. In addition to these regulatory fines, companies also risk losing customer loyalty and eventually destroying the brand equity that has built over a long time.

Businesses rely on data to support daily business operations and also to innovate and serve their customers more effectively, so it is essential to ensure privacy and protect all data, regardless of where it resides. Here are some key things you must consider to accomplish data privacy:

1. Data discovery: Understand what data exists across the different repositories.
2. Protect sensitive data: Leverage both encryption and data obfuscation for structured and unstructured data.

3. Protect non-production environments: Data copies are made for quality assessments and testing applications through a subset of production datasets. These datasets need to be obfuscated while retaining the structure and integrity necessary for effective testing.

4. Secure and monitor data access: Enterprise databases require real-time monitoring to ensure the data is protected, only the right people have access to it, and that all the interactions with data are audited.

With Cloud Pak for Data, the data privacy service allows you to define data protection rules and automated policy enforcement to ensure only authorized users have access to the data, as well as to ensure all protected data attributes are obfuscated automatically, as defined by the data protection rules:

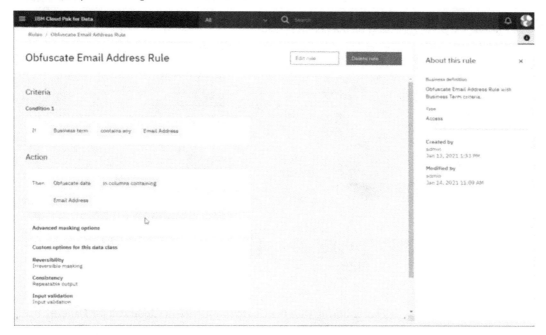

Figure 4.19 – Data rules for obfuscating data

Non-production applications often require access to representative datasets for testing. As a recognized best practice, de-identifying data provides the most effective way to protect the privacy and support compliance initiatives in non-production environments. The capabilities for de-identifying confidential data must allow you to protect privacy while still providing the necessary realistic data for development, testing, training, and other legitimate business purposes. With Cloud Pak for Data, you can also generate representative data, thereby obfuscating sensitive data attributes for all the related data to preserve application/data integrity.

Databases contain an organization's most valuable information, including customer data, payment card information, and financial results. Hackers are skilled at using techniques to penetrate the perimeter defenses and reach the databases. Traditional intrusion detection software does not understand the database protocols and structures required to detect inappropriate activities. What is required is real-time database activity monitoring to protect sensitive data and reduce business reputation risk.

Cloud Pak for Data's Guardium External S-TAP service works with your databases to perform compliance monitoring and data security. This service intercepts **TCP/IP** traffic either in plain text or in an encrypted format by parsing the activities, automatically enforcing predefined policies, and logging activity for further reporting:

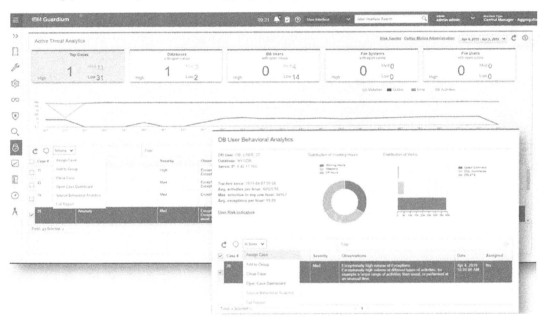

Figure 4.20 – Data activity monitoring with the Guardium service in Cloud Pak for Data

With that, we have gained a good understanding of the available data in the enterprise. Now, let's look at the data integration capabilities of the platform, which helps us bring together different datasets and manipulate the data so that it suits our needs.

Data integration at scale

Based on *IDC reports*, it is estimated that data scientists spend 80% of their time preparing and managing data for building AI models. This, coupled with an IBM survey where 91% of organizations are not using their data effectively, means that businesses are struggling to deliver value from data silos. Data integration is a combination of technical and business processes that are used to combine data from disparate sources into meaningful and valuable information. This is a critical aspect of an organization's overall information architecture strategy:

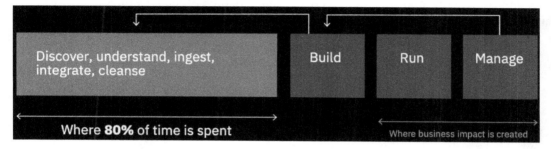

Figure 4.21 – Data integration, an enterprise challenge

IBM's DataStage service on IBM Cloud Pak for Data delivers a modernized data integration solution for cleansing and delivering trusted data anywhere, at any scale and complexity, on and across multi-cloud and hybrid cloud environments. There are many reasons for this major shift in how data integration tools are deployed and used with the rise of AI. The increase in volume and variety of data is one, but organizations are seeing a need to adopt a process-oriented approach to manage the data life cycle with DataOps, improve business performance, and increase competitiveness.

Considerations for selecting a data integration tool

The following are some key considerations we should look at when selecting a data integration tool/service:

- The ability to cater to a multi-cloud environment
- Performance at scale
- Integration with governance catalogs
- A wide array of connectors that support traditional RDBMS sources and other non-traditional sources of data
- The ability to track lineage
- Integrated data quality

With these considerations in mind, let's look at how we can use CP4D services to extract data from different sources, apply transformations, and load the results to different targets.

The extract, transform, and load (ETL) service in Cloud Pak for Data

Businesses that are embracing AI for their products and processes require a highly flexible and scalable data integration tool. IBM DataStage from IBM Cloud Pak for Data is equipped with the following features, which improve the productivity of your business and IT users:

- A best-in-class parallel engine and automatic workload balancing to elastically scale workloads

- The ability to design once, run anywhere, to bring data integration to where the data resides

- Automated job design and seamless integration with other services on the platform

- Increased user productivity with built-in design accelerators such as stage suggestions, schema propagation, and job template generation:

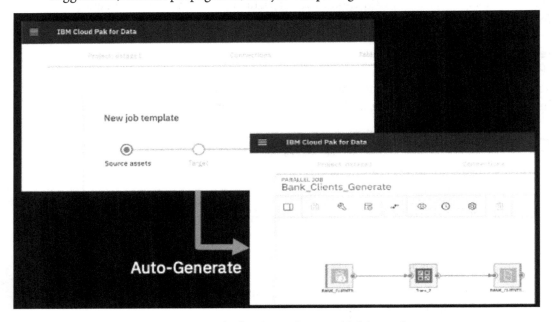

Figure 4.22 – Standardization and reuse with job templates

Now, let's understand how a cloud-native platform can help with data transformations.

Advantages of leveraging a cloud-native platform for ETL

From an architectural standpoint, IBM Cloud Pak for Data brings data integration to cloud-native platforms by breaking down the monolithic stack into microservices. Some of the advantages of taking this modernized approach include the following:

- Being able to deploy in minutes, enabling standard deployment and management while retaining the flexibility to modify environment/job parameters as required.

- Increased reliability due to inherent high availability in Kubernetes/OpenShift.

- Reduced management burdens due to automated updates.

- You can automate management by "application group," so administrators can use namespaces to manage access control and provisioning options.

- You can monitor and manage at the application level with platform- and service-level features.

- You can scale microservices independently to respond to changing needs:

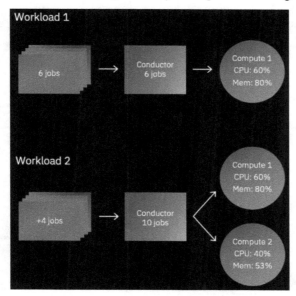

Figure 4.23 – Automatic workload balancing

Containerizing your data integration technology enables you to operate in a hybrid cloud environment across on-premises and cloud platforms.

Master data management

Organizations have a lot of data surrounding products, services, customers, and transactions, and it is important to derive insights from all of this data to maintain a competitive advantage. Master data management is a comprehensive process that defines and manages an organization's critical customer and product data. It provides a single, trusted view into data across the enterprise, with agile self-service access, analytical graph-based exploration, governance, and a user-friendly dashboard to accelerate product information management and master data programs.

Extending MDM toward a Digital Twin

A Digital Twin is a generic concept that was originally derived from the **Internet of Things (IoT)** space, wherein we have a digital representation of a physical thing. In the context of MDM, this translates to the ability to capture user interactions and derived attributes that can help service the customers more effectively, using the information that we know about them. This takes us beyond the standard attributes such as name, address, contact details, product details, and others we would source from MDM:

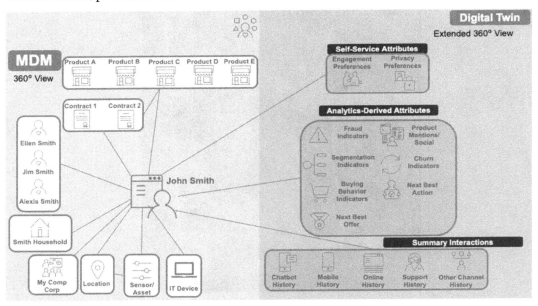

Figure 4.24 – Extending MDM toward Digital Twin

With Cloud Pak for Data's matching service, we can make this a reality by using a data-driven approach and reducing the time to value from months to hours/days, wherein the user can provision the service and start consuming the data in a very short amount of time. The platform connections enable access to various sources of data, and we can also leverage other platform capabilities such as the Watson Knowledge Catalog service to profile data, classify data, and perform machine learning-based mapping for MDM attributes:

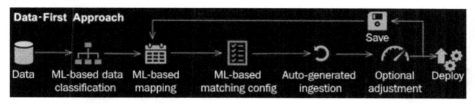

Figure 4.25 – Data-first approach

Data first is a methodology we can use to extract more value from data. With it, you can understand where you are today, and what the next best step to take is.

We can enable this capability with two different user experiences:

- The data engineer experience, which primarily focuses on configuring the master data.

- The line of business user experience, which is used to leverage the results of the matching service and uses entity information for analytics.

This is a three-step process to value:

1. The Data Engineer would either define a new model or use an existing MDM model, intending to extend it. As part of this process, you must decide what record types you want to master and what relationships will deliver insights.

2. The Data Engineer loads data from the source systems, profiles the data as part of the process, and maps the source attributes to the target attributes.

3. The data scientists and business analysts use the matching results for search and analytics purposes.

There are some out-of-the-box record types, such as Person. The service extends the capability for custom resource types to be added. The attributes of these resource types could either be simple attributes or complex attributes that comprise two or more attributes. The Data Engineer has the flexibility to take an interactive approach to building out the model over time, and then publishing the model back to MDM when the changes have been made.

Once the model has been defined, mapping the source attributes to the attributes in the model can be done by manually mapping them or using the ML-powered process of automatically classifying the attributes. As part of this step, it is possible to also eliminate redundant attributes and/or attributes that will not be mapped back to the MDM attributes:

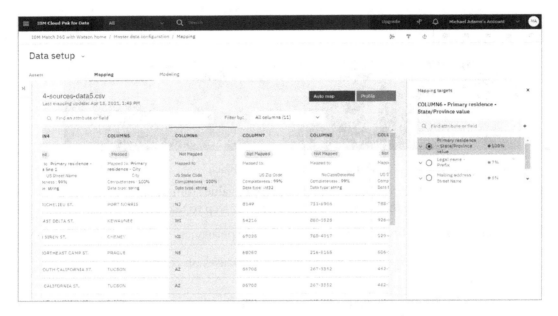

Figure 4.26 – Attribute, manual, and automated matching

Once the matching process is complete, the end users can search for Person and/or other custom record types and find results either by entities or individual record types. Let's use a simple example to understand the value of entity information. Every organization has customer profile data, and they also track the prospects of future business opportunities. Business users can benefit from entity data by getting answers to the following questions:

- Does the prospect list have duplicates?

- How many existing customers are on the prospect list?

- Are there any non-obvious relationships between the customers and prospects?

- Can I get better data points from new sources of information?

Data scientists can now understand the de-duplicated datasets either in a tabular format or by visualizing them using a relationship graph. This data can also be downloaded.

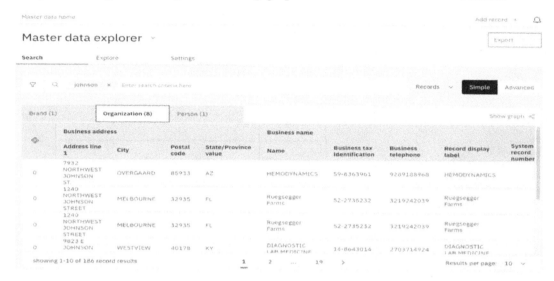

Figure 4.27 – Search and exploration

Let's sum up all that we have learned so far with the Summary next.

Data scientists can now understand the de-duplicated datasets either in a tabular format or by visualizing them using a relationship graph. This data can also be downloaded.

Summary

We started this chapter with the objective of accomplishing the Know -> Trust -> Use goal for all enterprise data. We now understand that DataOps is a process that leverages automation for handling data and gaining an understanding of data before it is shared through a centralized governance catalog, making it available to be searched for and used by the broader organization. We saw how the various services in the IBM Cloud Pak for Data platform play a role in profiling, classifying, and manipulating data to create a trusted data platform that fosters collaboration across the enterprise.

In the next chapter, we will review the analytic needs of an enterprise and how the analysis capabilities in the platform address the different personas' needs.

5
Analyzing: Building, Deploying, and Scaling Models with Trust and Transparency

There is a lot of data collected across an enterprise, and this presents a unique opportunity to harvest insights that can be used to improve efficiency and serve our customers better. **IBM Cloud Pak for Data** has an extensive collection of services from **International Business Machines Corporation (IBM)**, as well as third-party services that help with the various aspects of analyzing data.

The platform caters to different personas including business analysts, data engineers, data scientists, data stewards, and so on, enabling cross-persona collaboration. All of these personas are interested in different aspects of data and they have varied skillsets when it comes to analyzing the data, starting from something as simple as the need to execute canned reports on a regular basis to ad hoc reporting and dashboarding and more advanced analytics such as statistical analysis, **machine learning** (**ML**), or **artificial intelligence** (**AI**).

Enterprises often have very diverse analytics/data science teams, from members that are clickers who consume data generated by others and create reports/dashboards to more advanced coders who understand the math and are able to either create or leverage existing algorithms and/or techniques to complete their analytics tasks.

In this chapter, we will start with defining the different types of analytics and key considerations for analytics in a data and AI platform, and then look at how the platform services can be leveraged to address the needs of an enterprise analytics team.

We will cover the following topics in this chapter:

- Self-service analytics of governed data
- **Business intelligence** (**BI**) and reporting
- Predictive versus prescriptive analytics
- Understanding AI
- AI life cycle: Transforming insights into action
- AI governance: Trust and transparency
- Automating the AI life cycle using Cloud Pak for Data

Self-service analytics of governed data

All enterprises have data and BI teams across the different business units. The traditional process for a business user to get reports would be that they submit a request for one or more reports across different subject areas to these data and BI teams. The data team would identify the best source(s) of data, construct a query based on the business requirement, and do an initial validation of the query results. This is then passed on to the BI team for report creation, after which the business user gets the report they are looking for. This can often be an iterative process and can take anywhere from days to weeks.

Self-service analytics is the ability of a **line of business** (**LOB**) to generate reports and dashboards without reliance on **Information Technology** (**IT**). Self-service analytics helps business analysts and data analysts bring together data from different sources, perform queries, and create reports that help make important business decisions. This brings agility to analytics; however, it can also potentially introduce risk if it lacks proper governance measures.

Cloud Pak for Data, with its centralized governance catalog and dashboarding service, enables self-service analytics and the confident sourcing of data from governed data assets.

BI and reporting

BI is a capability that ingests business data and presents it in user-friendly views such as reports, dashboards, charts, and graphs. BI tools enable business users to access different types of data, including historical and current, third-party, and in-house, as well as semi-structured data and unstructured data such as social media. Users can analyze this information to gain insights into how the business is performing. Although BI does not tell business users what to do or what will happen if they take a certain course of action, neither is BI only about generating reports. Instead, BI offers a way for people to examine data to understand trends and derive insights.

Predictive versus prescriptive analytics

Predictive analytics is a practice of using advanced algorithms and ML to process historical data, learning what has happened while uncovering unseen data patterns, interactions, and relationships. Predictive models provide actionable insights, but they don't say what action needs to be taken based on insights for best outcomes.

Prescriptive analytics, on the other hand, enable accurate decision-making for complex problems by using optimization models that are mathematical representations of business problems. These optimization models use solvers with sophisticated algorithms.

Often, we might question whether we should use predictive analytics or prescriptive analytics. The answer to that is we need both, and to illustrate the point, let's use a simple example. Throughout the Covid-19 pandemic, ML models have been used to forecast the demand for protective masks, but complementing this information with prescriptive analytics will help build replenishment plans to decide which distribution center should supply to different locations to adequately meet demand.

The following diagram illustrates the different types of analytics and the questions we can answer by leveraging each of these types:

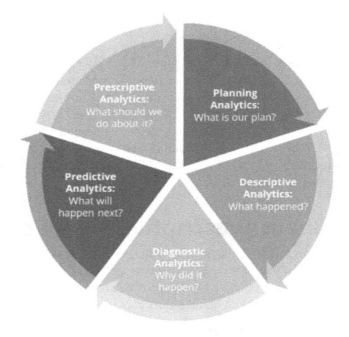

Figure 5.1 – Types of analytics

Now we have a good understanding of the different types of analytics, let's define AI and how we transform insights into action.

Understanding AI

AI refers to a human-like intelligence exhibited by a computer, robot, or other machines. It is the ability of a machine to mimic the capabilities of a human mind, learning from examples and experience, recognizing objects, understanding and responding to language, and using all these data points to make better decisions and help solve real-world problems.

AI is increasingly becoming part of our everyday lives, and this trend is only going to grow over the next few years. We can see AI in action, completing our words as we type them, providing driving directions to the nearest restaurant, recommending what we can eat, and even completing household chores such as vacuuming the house, and so on. As popular as it is, there is often some confusion in its terminology, and there is a tendency to use AI, ML, and **deep learning** (**DL**) interchangeably.

Let's start by demystifying these three concepts, as follows:

- **AI**—AI systems can be thought of as expert systems that can connect different data points and make decisions based on complex rules that mimic human behavior, free of emotions. AI is used to predict, automate, and optimize tasks that have historically been done by humans, such as speech recognition, facial recognition, translation, and so on.

- **ML**—A subset of AI applications that learns by itself, using new data points to predict outcomes with increasing accuracy.

- **DL**—A subset of ML where neural networks learn from large volumes of data. DL algorithms perform a task repeatedly and gradually improve the outcome through deep layers that enable progressive learning.

The following diagram shows the relationship between AI, ML, and DL:

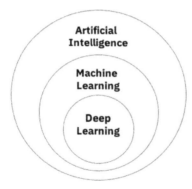

Figure 5.2 – Types of AI

With a better understanding of AI, ML, and DL, we will now look at the existing methodology for ML projects and additional considerations for new-age AI applications to transform insights into action.

AI life cycle – Transforming insights into action

Enterprises going through digital transformation infuse AI into their applications; a well-defined and robust methodology is required to manage the AI pipeline. Traditionally, a **cross-industry standard process for data mining** (**CRISP-DM**) methodology was used for ML projects, and it is important to understand this methodology before we explore the challenges faced with the implementation of an AI-driven application and how a more comprehensive AI life cycle will help with this process.

CRISP-DM is an open standard process model that describes the approach and the steps involved in executing data mining projects. It can be broken down into six major phases, as follows:

- **Business understanding**
- **Data understanding**
- **Data preparation**
- **Model building**
- **Model evaluation**
- **Model deployment**

The sequence of these phases is not strict and is often an iterative process. The arrows in the following diagram indicate the most important and frequent dependencies between the different phases, while the outer circle symbolizes the cyclic nature of data mining:

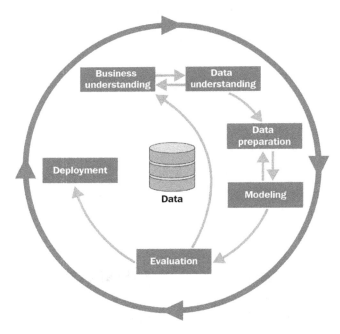

Figure 5.3 – Analytics life cycle

This worked great in the past, but with the increasing amount of data being collected, we now have to consider the volume, variety, and velocity of data and how it affects the AI life cycle. What this means is we need to place more emphasis on the data collection, data governance, and continued model evaluation phases. Model evaluation is no longer just about evaluating the quality of the model—there are also other considerations, including the fairness and explainability of the models being deployed.

Here are some key considerations when enabling this AI life cycle:

- Lowering the skillset to build and operationalize AI
- Accelerating delivery time by minimizing mundane tasks
- Increased readiness for AI-infused applications through optimizing model accuracy
- Delivering real-time governance of AI models, improving trust and transparency

With a good understanding of the additional considerations for AI applications, we will next delve into the need for AI governance as a foundation of trust and transparency in new-age AI applications.

AI governance: Trust and transparency

AI is prevalent in every industry—everything from recommendation engines to tracking the spread of the COVID-19 virus. This application of AI empowers organizations to innovate and introduce efficiency, but it also introduces risks around how AI is employed to make business decisions.

In general, governance usually relates to who is responsible for data, how data is handled, who has access to see particular datasets in an enterprise, and how they get to use the data. This same notion of establishing a governance framework to protect the data and the consumers also applies to AI models, which may exhibit behavior that may be unfair or biased—or both—to some individuals or groups of people. Data governance has evolved over a period, and there are well-established methodologies and best practices, but we now need to do something similar for AI models. The challenge with traditional data governance is the increasing volume of data and the need to identify and align data/technical assets to business assets that are organized into subjects and logical areas and establishing policies and rules that dictate the use of data assets. However, with AI models, there are several considerations, including the following ones:

- Multiple versions of the model
- Different input datasets used to train the model
- Pipeline information on any/all transformations to the model input attributes

- Algorithm/technique used to train the model

- Hyperparameter optimization

- Training/test datasets used to build and test the model

Explainable AI is not just about being able to interpret the model output—it is also about being able to explain how the model was built, which input data selection process and filters were used, whether the selected dataset was in compliance with organizational policies and rules, and if there was any intentional or unintended bias in the selected datasets.

Automating the AI life cycle using Cloud Pak for Data

After an organization has been able to collect its data and organize it using a trusted governance catalog, it can now tap into the data to build and scale AI models across the business. To build AI models from the ground up and scale it across the business, organizations need capabilities covering the full AI life cycle, and this includes the following:

- **Build**—This is where companies build their AI models.

- **Run**—After a model has been built, it needs to be put into production within an application or a business process.

- **Manage**—After a model is built and running, the question becomes: *How can it be scaled with trust and transparency?* To address complex build and run environments, enterprises need a tool that not only manages the environment but also explains how their models arrived at their predictions.

Let's exercise **Model Operations** (**ModelOps**) using Cloud Pak for Data to understand how the platform can be used to automate the AI life cycle. ModelOps is a methodology and process of building data science assets that can quickly progress from development to production.

We will use a simple use case of predicting customer churn to walk through the different phases of ModelOps. The input dataset contains customer demographics, account information, and whether they are churned or not. We will use the different services in Cloud Pak for Data to source this input data, do some data transformations, build ML models, deploy the models, and operationalize them.

The following diagram illustrates the efficiencies we have gained through DevOps, resulting in the collection of more data, now presenting an opportunity to bring more efficiencies to ModelOps by automating the different steps in the process:

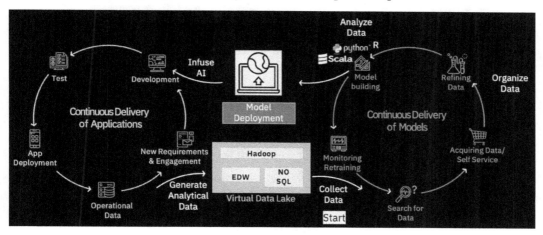

Figure 5.4 – ModelOps

Over the past decade, organizations have modernized their applications and introduced agility through the use of **Development and Operations (DevOps)**, which has enabled fast and continuous delivery of changes to applications to cater to changing business needs. What we are going to accomplish with ModelOps will bring the same agility, to enable the continuous delivery of models.

We will use the following Cloud Pak for Data services to walk through a scenario to demonstrate automation of the AI life cycle and contrast it with the traditional way of doing things.

The following screenshot describes the overall process and the interactions between data engineers and data scientists in a traditional setup versus the *additional capabilities* enabled through ModelOps in Cloud Pak for Data:

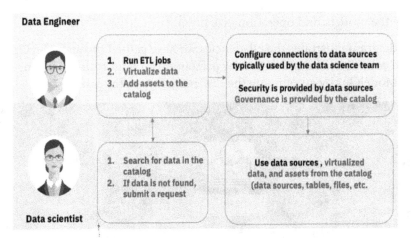

Figure 5.5 – Process improvements with ModelOps

With an understanding of the efficiencies we get with ModelOps, let's now explore the needs of a diverse data science team in terms of the tools/technologies they need to leverage in their day-to-day jobs.

Data science tools for a diverse data science team

Today's enterprise has a diverse data science team with clickers (someone who is comfortable using a drag/drop interface to define a pipeline as opposed to writing code to achieve the same result) and coders that demand a choice of tools and runtimes through a combination of open source and commercial technologies. These are some of the services included in Cloud Pak for Data:

- **Open source services: Jupyter, Jupyter Notebook, RStudio**

- **Service for clickers**: Modeler flows

- **Automated AI (AutoAI)**: Helps automatically create classification and regression models

Data scientists and developers are a rare commodity these days, and their time is invaluable. IBM Cloud Pak for Data addresses this challenge through its industry-leading AutoAI capability in the Watson Studio service, which enables citizen data scientists to build, deploy, and manage AI models. By infusing automation and democratizing AI, IBM expands the reach of AI to business users. Furthermore, AutoAI complements data scientists and developers by automating some of the more mundane tasks of the AI life cycle such as data preparation, model selection, feature engineering, and hyperparameter optimization. Finally, models built and deployed through AutoAI can be updated by data scientists, unlike with other vendors, where any automation is a black-box one.

The following diagram shows the different steps involved in training one or more AI models and selecting the best model. AutoAI brings automation to all these steps, as well as the convenience of testing multiple algorithms:

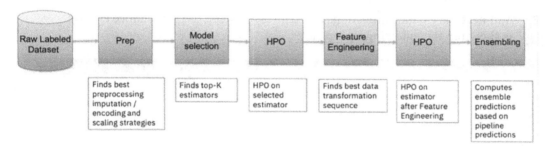

Figure 5.6 – Model training and selection process

After covering the data science tools for a diverse data science team, we'll now cover distributed AI.

Distributed AI

We live in a complex world with data and teams distributed across multiple geographies, multiple data stores, and—in some cases—multiple clouds (private and public). Building enterprise AI models with this complexity is a tremendous challenge, but IBM Watson Studio includes everything a customer needs to address this challenge. IBM Cloud Pak for Data supports building models across distributed datasets through **federated learning (FL)**. This helps minimize data movement and, more importantly, enables enterprises and consortiums to build centralized models without sharing data. Additionally, collaboration is infused throughout, ensuring data scientists can work on projects as one team. Native Git integration enables the versioning of models, a critical requirement for enterprise AI.

Establishing a collaborative environment and building AI models

The first step in the process is to establish a project. A Cloud Pak for Data analytics project is used to organize assets and collaborate across the organization. The tools available for this project depend on which services are enabled on the platform.

Analytics projects can either be created from scratch or we can leverage an import mechanism to import existing projects. Once a project is created, we go through the following steps:

1. Add collaborators to the project.

2. Customize any runtime environments as required. There are some existing runtime environments that can be installed with the platform.

3. Add necessary data connections and data assets from the governance catalog, as shown in the following screenshot:

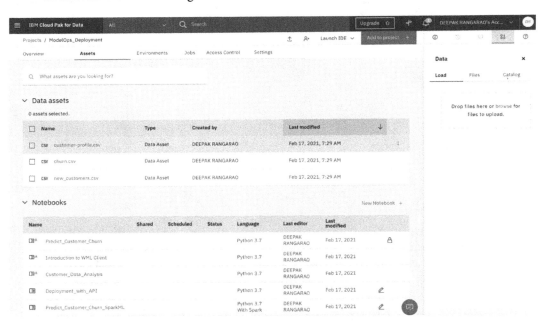

Figure 5.7 – Project-based collaborative development environment

Now we have established the environment, we will show how to choose the right tools in the following section.

Choosing the right tools to use

The Watson Studio service in Cloud Pak for Data allows you to choose from a wide range of tools in analytics projects for all types of users. These users can have varying levels of experience in preparing, analyzing, and modeling data, from beginners to expert users. To pick the right tools for the task, here are the key considerations:

- The type of data to work with, such as the following:

    Tabular data in delimited files or relational data sources

    Image files

    Textual data in documents

- The type of task that needs to be accomplished, such as the following:

    Prepare data: Cleanse, shape, visualize, organize, and validate data

    Analyze data: Identify patterns and relationships in data, and display insights

    Build models: Build, train, test, and deploy models to classify data, make predictions, or optimize decisions

- The level of automation desired, such as the following:

    Code editor tools: Use these to write code in `Python`, `R`, or `Scala`

    Graphical canvas tools: Use a drag-drop interface to visually define a pipeline

    Automated tools: Use to configure automated tasks that require limited user input

With the preceding considerations, the following table helps pick the best tools for the task at hand:

Tool	Tool type	Prepare data	Analyze data	Build models
Jupyter Notebook	Code editor	✓	✓	✓
JupyterLab	Code editor	✓	✓	✓
RStudio	Code editor	✓	✓	✓
Data Refinery	Graphical canvas	✓	✓	
Streams flow	Graphical canvas	✓	✓	
Dashboard	Graphical canvas		✓	
Decision Optimization model builder	Graphical canvas, code editor	✓		✓
AutoAI	Automated	✓		✓
Metadata import	Automated	✓		
Statistical Package for the Social Sciences (SPSS) modeler	Graphical canvas	✓	✓	✓

Figure 5.8 – AI/ML tools aligned to different steps in the AI model building process

The following table shows available tools for building a model that classifies textual data:

Tool	Code editor	Graphical canvas	Automated tool
Jupyter Notebook	✓		
JupyterLab	✓		
RStudio	✓		
SPSS modeler		✓	
Experiment builder		✓	

Figure 5.9 – Tools available for textual data classification

The following table shows tools for image data:

Tool	Code editor	Graphical canvas	Automated tool
Jupyter Notebook	✓		
JupyterLab	✓		
RStudio	✓		
Experiment builder		✓	

Figure 5.10 – Tools available for analytics on image data

We have seen the tools/technologies to handle different types of data. Let's now look into the process of establishing a collaborative development environment.

Establishing a collaboration environment

In the building-AI-models phase of the AI life cycle, data scientists need to collaborate with data engineers and other data scientists to build the AI models, which could involve sharing data assets and/or existing data pipelines. For a given use case, data scientists train tens or hundreds of AI models before picking the best model. To support and maximize the productivity of the data science team, Cloud Pak for Data supports multiple modes of collaboration, including the following:

- Local project collaboration without Git
- Collaboration via Git

A local collaboration model enables all collaborators to work on one copy of assets in a project. Any asset being updated by one user (for example, a notebook) is locked until that user or an administrator unlocks it. This enables changes to be immediately available to all collaborators.

The following diagram shows a collaboration model using project-level authorization with the Watson Studio local service in Cloud Pak for Data:

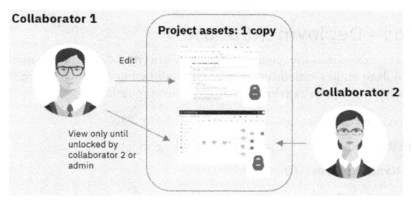

Figure 5.11 – Collaboration model with Cloud Pak for Data

Collaboration via Git is enabled when a project is connected to a Git repository. Once a project is connected, users can use the standard Git mechanism of commit, push, and pull to synchronize the assets in the project. All changes will be reflected in the project when they are pushed to the Git repository by the editor. Changes can also be tracked in the Git commit history.

The following diagram illustrates **continuous integration/continuous delivery (CI/CD)** in the context of Cloud Pak for Data:

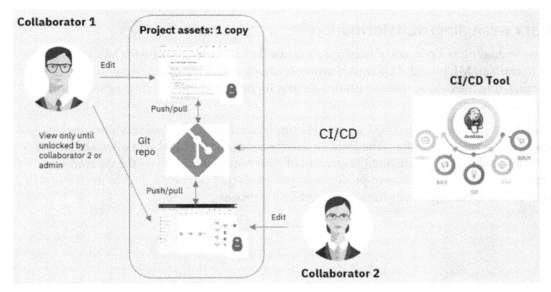

Figure 5.12 – CI/CD with Cloud Pak for Data

Upon establishing a collaborative development environment, the next step is to understand the different steps involved in the deployment phase of ModelOps.

ModelOps – Deployment phase

Once a project is created and the required data assets are created, we can use one of the available tools to build a classification model for predicting customer churn. In this example, we will use a Jupyter notebook to do the following tasks:

- Access training data for customer churn
- Explore the input dataset using a pandas profiling widget
- Train a **Random Forest** (**RF**) model
- Save the trained model

The remaining parts of this section will walk you through the considerations for the deployment phase of ModelOps.

Trusted data access

The quality of AI models is dependent upon the input dataset used, so it is important we use data we can trust. The centralized governance catalog in Cloud Pak for Data helps set up a governance framework and helps end users search and consume high-quality data. In this case, we will use directly source data from the catalog in the Jupyter notebook.

Data wrangling/transformation

Data wrangling is a process of manipulating raw data to make it useful for analytics or to train an ML model. This could involve sourcing data from one or more sources, normalizing the data so that the unified dataset is consistent, and parsing data into some structure for further use.

Cloud Pak for Data has a choice of data manipulation/transformation services suitable for both clickers and coders. The Data Refinery service has an Excel-like interface for data wrangling, drastically reducing the amount of time required to prepare the dataset. This service has a large collection of predefined operations that can be used to define transform operations as multiple steps in the data-wrangling process.

For data scientists that have experience with programming/scripting languages, they can choose to use the Jupyter notebook interface in conjunction with different runtimes to programmatically do the data transformation.

The following screenshot shows an Excel-like interface for data wrangling in Cloud Pak for Data:

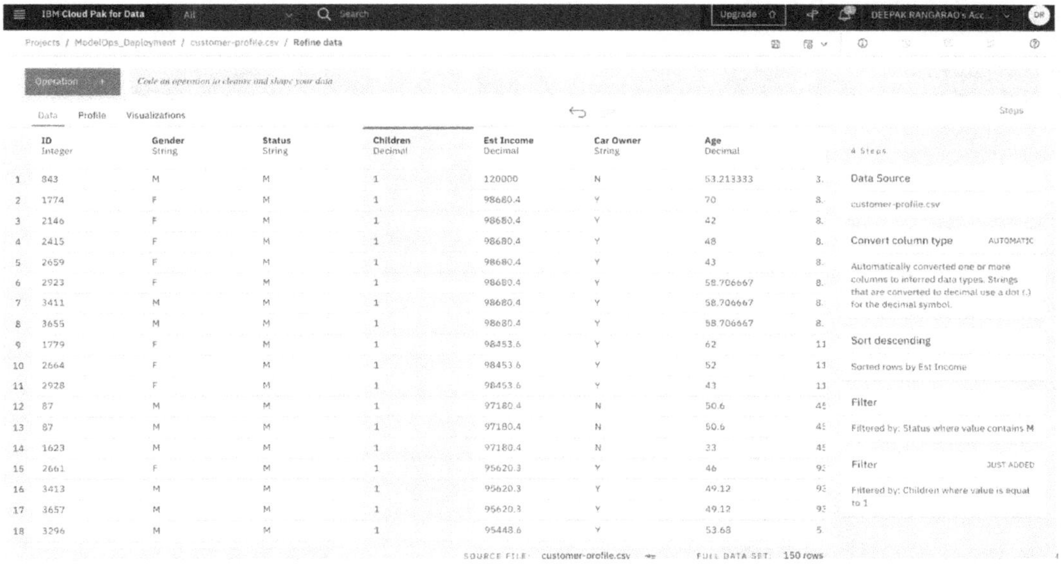

Figure 5.13 – Data wrangling for non-Extract, Transform, Load (ETL) developers

Once the data is structured in the desired format, the next step is to use this data and train models to help predict target variables.

Model building

Once the data is transformed and you have it in the desired shape, the next step is to decide which algorithm to use and build a model with it. You have the choice to either use a programming interface or a drag-drop **user interface** (**UI**). Cloud Pak for Data supports Jupyter Notebooks, JupyterLab notebooks, and the RStudio interface to leverage existing libraries/modules/**Comprehensive R Archive Network** (**CRAN**) packages and build ML models.

Cloud Pak for Data also includes the SPSS modeler flow service, a drag-and-drop interface to source data, apply transformations, train, and score ML models. There is a wide range of palette nodes, ranging from import operations to field/record operations, modeling, and visualization.

The following screenshot shows the drag-and-drop interface for data preparation, statistical analysis, model building, inference, and visualization:

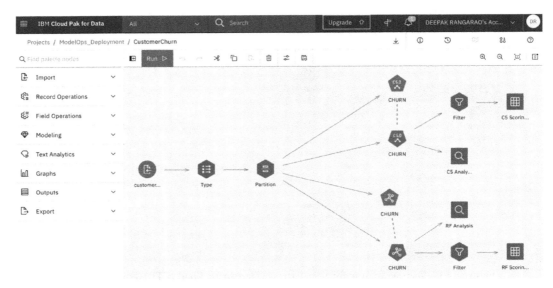

Figure 5.14 – SPSS modeler flows for visual model building

Now we have covered the steps to use data and train models to help predict target variables, we'll be walking you through the accelerators for the Jupyter Notebook interface.

Accelerators for the Jupyter Notebook interface

While Jupyter Notebook is open source software, Cloud Pak for Data has enhanced the experience of using the Jupyter Notebook interface by adding enterprise capabilities such as Git integration and also accelerators such as code generation for assets in the catalog.

The following screenshot illustrates the Cloud Pak for Data usability enhancements to the open source Jupyter Notebook interface:

Figure 5.15 – Code generation with Watson Studio local notebooks

We have seen how to do data wrangling and train models using the available tools/ technologies. The next step is to understand the model deployment process and options available in the platform.

Model deployment

Once the model is saved to the repository, we will then go through the process of promoting the model to the deployment space associated with the project, configuring the deployment, testing the deployment, and making it available for integration with applications.

Cloud Pak for Data supports three deployment options, outlined as follows:

- **Online**: A real-time request/response deployment option. When this deployment option is used, models or functions are made available for invocation using a **Representational State Transfer (REST) application programming interface (API)**. A single row or multiple rows of data can be passed in with the REST request.

- **Batch**: This deployment option enables reads and writes from/to a static data source/target and can also be invoked via a REST API.

- **CoreML**: This deployment option provides the capability to save a model for deployment in CoreML Runtime. CoreML is not included with Cloud Pak for Data.

  Here is a screenshot of the deployment models in the Watson Machine Learning service:

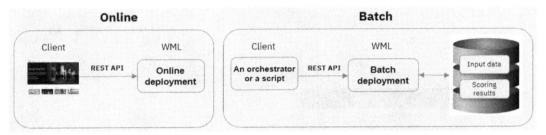

Figure 5.16 – Deployment models with the Watson Machine Learning service

In addition to models, it is also possible to deploy other assets, including the following:

- Notebooks as read-only **HyperText Markup Language** (**HTML**) pages

- Notebooks as a scheduled batch job for automation

The process of promoting a model to a deployment space can be either via an API or from the UI. The API interface enables deployment from multiple client interfaces, both the notebook interface available in Cloud Pak for Data and external client tools such as PyCharm, **Visual Studio Code** (**VS Code**), and so on. The deployment flow for most asset types includes the following steps:

1. Create a deployment space.

2. Save the asset (model, **Predictive Model Markup Language** (**PMML**), and so on) to the project repository.

3. Promote the asset to the deployment space.

4. Configure the deployment (online or batch).

5. Integrate the deployed asset with other applications via a REST API.

While configuring deployments, we have the flexibility to not just define them as online or batch model deployments but also to specify the hardware specification for the actual deployment. Once the model is deployed, developers have access to the **software development kit (SDK)** in multiple languages to score against the model and operationalize it. The UI also allows for the testing of deployment models by passing through the input payload.

The following screenshot illustrates the testing of deployment models with the Watson Machine Learning service in Cloud Pak for Data:

Figure 5.17 – Model testing with Watson Machine Learning

Now we have covered model deployment, let's explore the ModelOps monitoring phase.

ModelOps – Monitoring phase

Building and deploying models is only a small sliver of the AI life cycle. Managing the deployed models and ensuring that they are accurate and compliant with regulatory requirements such as **System Requirement (SR)** *11-7* is a whole different ball game. IBM Cloud Pak for Data comes with a wide array of built-in capabilities to ensure trust and transparency. This includes governance of training data and AI models to ensure compliance, such as model explainability, fact sheets to capture model metadata and lineage for auditing, model monitoring, and automatic training of models across IBM and non-IBM deployments. As more and more enterprises operationalize AI, infusing governance and managing models at scale with trust and transparency is at the forefront, and there is no one better than IBM to deliver this.

As businesses go through digital transformations and modernize applications by infusing AI, there is a need for more visibility into the recommendations made by AI models. In some industries such as finance and healthcare, there are regulations such as the **General Data Protection Regulation (GDPR)** and others that present significant barriers to widespread AI adoption. Applications must have the ability to explain their outcomes in order to be used in production systems. Every decision made as a result of incorporating AI needs to be traceable, enabling enterprises to audit the lineage of models and the associated training data, along with the **inputs and outputs (I/Os)** for each AI-based recommendation.

The OpenScale service in IBM Cloud Pak for Data enables businesses to operate and automate AI at scale, irrespective of how the AI system was built and where it runs, bridging the gap between the teams that operate AI systems and those that manage business applications. Businesses now have confidence in AI decisions as they are now explainable.

The Cloud Pak for Data OpenScale service supports the following:

- AI deployed in any runtime environment (for example, **Amazon Machine Learning (Amazon ML)**, **Azure Machine Learning (Azure ML)**, and custom runtime environments, behind the enterprise firewall)
- Applications and ML and DL models developed using any open source model building and training **integrated development environment (IDE)**, including TensorFlow, scikit-learn, Keras, SparkML, and PMML

The following screenshot illustrates model monitoring capabilities with the Watson OpenScale service in Cloud Pak for Data:

Figure 5.18 – Model monitoring with Watson OpenScale

Let's now move on to detecting AI bias.

Detecting AI bias

AI brings tremendous power and value to business operations, but it also introduces the potential for unintended bias. Throughout the AI life cycle, from model building to operationalization, this bias can change. As an example, we might use a training dataset that has an equal number of observations for male and female customers, so the model might not have any bias. Once we go to the production phase, we continue to retrain the model on a regular basis, and this might introduce an unintended bias.

It is important to monitor for bias at runtime and notify proactively so that we can act on it. OpenScale monitors the data that has been sent to the model and the model predictions. It then identifies bias issues and notifies the user of these. Let's review some core concepts before we understand how bias can be dealt with, as follows:

- **Fairness attribute**: Bias is generally measured in the context of one attribute—for example, gender, age, and so on. This can also be non-demographic attributes such as account information, how long a person has been a customer, and so on.

- **Monitored/reference group**: This is a subset of the dataset for which we want to measure bias. The remaining values for the fairness attribute can be called a reference group. If we want to measure gender bias toward males, females would be considered a reference group.

- **Disparate impact**: This is a measure of bias that can be computed as a ratio of the percentage of a favorable outcome for the monitored group to the percentage of a favorable outcome for the reference group. Bias is said to exist if the disparate impact value is below some predefined threshold.

- **Bias threshold**: If the bias threshold is 1, then it would mean that we expect both monitored and reference groups to get the same outcome of predictions.

The following screenshot illustrates the detection of bias with Watson OpenScale:

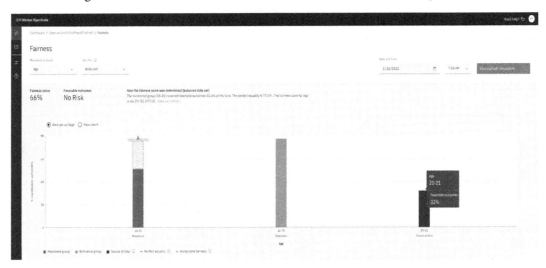

Figure 5.19 – Bias detection with Watson OpenScale

As we have now explored what detecting AI bias brings to business operations, let's next look into how data perturbation works and how the OpenScale service uses this technique as a differentiator to other options for bias detection.

Data perturbation

The OpenScale service uses a data perturbation technique as a differentiator to other options for bias detection. Let's use an example scenario of 100 insurance claims. In these claims, 60 claims were made by people in the *18-24* age group and 40 in the *above 24* age group. All the people in the *18-24* age group had very high claim frequency and were known to have made fraudulent claims in the past, hence the model rejected all 60 claims in that age group. On the other hand, the 40 claims in the *above 24* age group were all approved by the model. The disparate impact for this would be *0/1=0*, leading to the model being flagged as biased. However, the model is not biased and is based on the data provided to it. To avoid this situation, OpenScale leverages data perturbation.

Data perturbation works by flipping the record in the monitored group with a random value from the reference group and then sends the perturbed value to the model. All the other features of the record are kept the same. Now, after perturbation, if the model predicts that the claim in the preceding example is approved, then it means the model is acting in a biased manner by making decisions just based on the age of the customer. Thus, while computing bias, we do this data perturbation and compute the disparate impact using the payload and the perturbed data. This will ensure that we report bias only if there are enough data points where the model has exhibited biased behavior across the original and the perturbed data. Thus, we are able to detect if there is genuine bias in the model and not get impacted by the kind of data being received by the model.

Comparing model quality, fairness, and drift

The OpenScale service in Cloud Pak for Data enables you to monitor multiple models across different providers and algorithms, enabling customers to compare performance over a period of time and across different versions (production/pre-production).

The following screenshot illustrates the ability to compare the fairness and quality of models using Watson OpenScale:

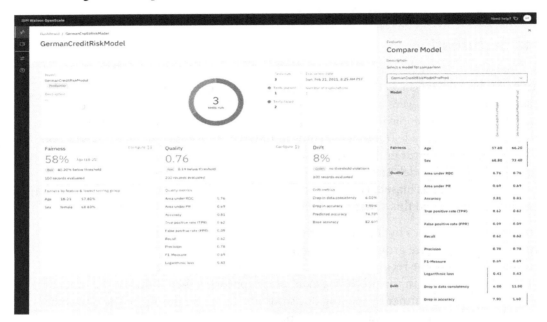

Figure 5.20 – Model comparison with Watson OpenScale

Having seen how to compare different models for fairness and quality, let's next explore how to use the platform capabilities to explain AI outcomes.

Explaining AI outcomes

Explainability is a very important capability from a regulatory standpoint. OpenScale provides the following two kinds of explanations for AI models:

- **Local interpretable model-agnostic explanations** (**LIME**)
- Proprietary algorithm for contrastive explanations

LIME treats all models as black boxes and generates an explanation that is easy to understand. However, this also requires that LIME understands the model behavior in the vicinity of the data points. This can lead to increased costs because of more scoring requests. The OpenScale service has an innovative caching-based technique that leads to a significant drop in the number of scoring requests required for generating a local explanation. In addition to reducing cost, there is also a significant improvement in the speed of generating an explanation. The OpenScale service has also added a fault tolerance feature to cater to intermittent failure in scoring requests.

The second explainability algorithm provides contrastive explanations. This algorithm is tailored to handle models that accept structured data and is able to generate contrastive explanations with far fewer scoring requests compared to open source alternatives.

Model drift

Model drift refers to a degradation in model performance due to changes in the underlying data and the relationship between I/O variables. Model drift can negatively impact over time, or sometimes there could be a sudden drift, causing undesirable model behavior. To effectively detect and mitigate drift, organizations need to monitor and manage model performance regularly as part of their data and AI platform. This integrated approach has the following benefits:

- Ability to track metrics on a continuous basis, with alerts on drifts in accuracy and data consistency.

- Ability to define targets and track them through development, validation, and deployment.

- Simplified steps to identify business metrics affected by model drift.

- Automating drift monitoring minimizes the impact of model degradation.

The OpenScale service detects drifts in accuracy and in the data. Estimating drifts in accuracy is done at model runtime. Model accuracy generally drops if there is an increase in transactions not covered by the model training dataset. This type of drift is calculated for structured binary and multi-class classification models only.

The following screenshot illustrates the ability to continually monitor models and evaluate model quality over time:

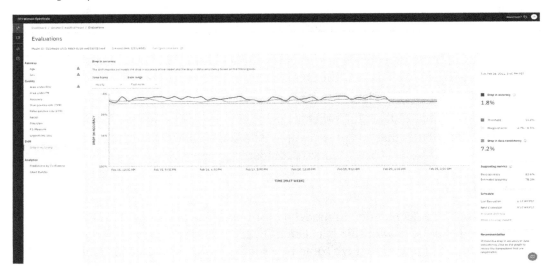

Figure 5.21 – Model evaluation over time

Now we have explored capabilities around predictive analytics, let's next look into how we can handle streaming data and perform analytics using the services in Cloud Pak for Data.

Streaming data/analytics

Data is growing at a rapid pace, and the variety and complexity of data are continually increasing. Every day, we create 2.5 quintillion bytes of data, and traditional analytics solutions are not able to fully unlock the potential value of that data. Cloud Pak for Data has the IBM Streams service to address streaming data needs. This includes an API, an IDE for applications, and a runtime system that can run the applications on a single or distributed set of resources. The Streams service is designed to address the following data processing platform objectives:

- Parallel and high-performance streams processing
- Connectors to a variety of relational databases, non-relational data sources, and multiple streaming data sources including IBM Event Streams, **HyperText Transfer Protocol (HTTP)**, and **Internet of Things (IoT)**
- Apply simple-to-advanced analytics methods, including ML, to real-time data streams
- Automated deployment of stream processing applications

- Incremental deployment without restarting to extend stream processing applications
- Secure and auditable runtime environment

There are several ways to develop applications that analyze streaming data, as outlined here:

- Streams flows from within a Cloud Pak for Data project
- Streams Python API from a Jupyter notebook in a Cloud Pak for Data project
- Streams Python API and any text editor or Python IDE

Now, let's move on to distributed processing.

Distributed processing

The Cloud Pak for Data service leverages the Red Hat OpenShift platform to schedule workloads from different services across the compute nodes in a cluster. There are some situations where we might either not have capacity in a cluster or want to run a workload at a remote location, either due to data locality or regulatory requirements that dictate where the workload can be executed geographically. Cloud Pak for Data has the following two options for distributed processing:

- The Execution Engine for Apache Hadoop service
- The Edge Analytics service

The Execution Engine for Apache Hadoop service enables you to create Hadoop environment definitions and **Jupyter Enterprise Gateway** (**JEG**) sessions from within Watson Studio analytics projects to run jobs remotely on the Hadoop cluster. The following tasks can be accomplished using this service:

- Train a model on a Hadoop cluster
- Manage a model on a Hadoop cluster
- Preview and refine Hadoop data (**Hadoop Distributed File System** (**HDFS**), Hive, and Impala)
- Run Data Refinery flows on the Hadoop cluster
- Schedule Python or R scripts as jobs to run on Hadoop clusters remotely

The following diagram illustrates distributed processing with Cloud Pak for Data using the Hadoop Integration service to push the processing closer to where the data resides:

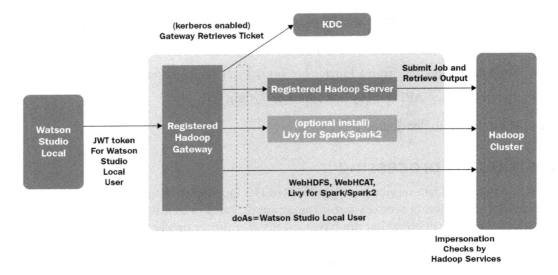

Figure 5.22 – Distributed processing with Hadoop Integration

The Edge Analytics service in Cloud Pak for Data enables the following use cases:

- **Faster insights and actions**: When there is a need for detecting situations and taking action in real time, a centralized analytics approach might not be the most effective as it adds too much network latency for immediate time-critical action to be taken.

- **Continuous operations**: Moving analytics to the edge, close to the device, also enables support for devices that must continue to provide critical functions even if disconnected from the **wide-area network** (**WAN**)/data center/cloud.

- **Better data and cost control**: With a massive growth in data volumes generated at the edge, outside the traditional data center there is a need to cost-effectively collect and analyze that data. Edge Analytics allows for filtering data at the source, aggregating data at the source where a finer level of detail is not required, and processing sensitive data at the edge.

The following diagram illustrates the distribution of work using the Edge Analytics service:

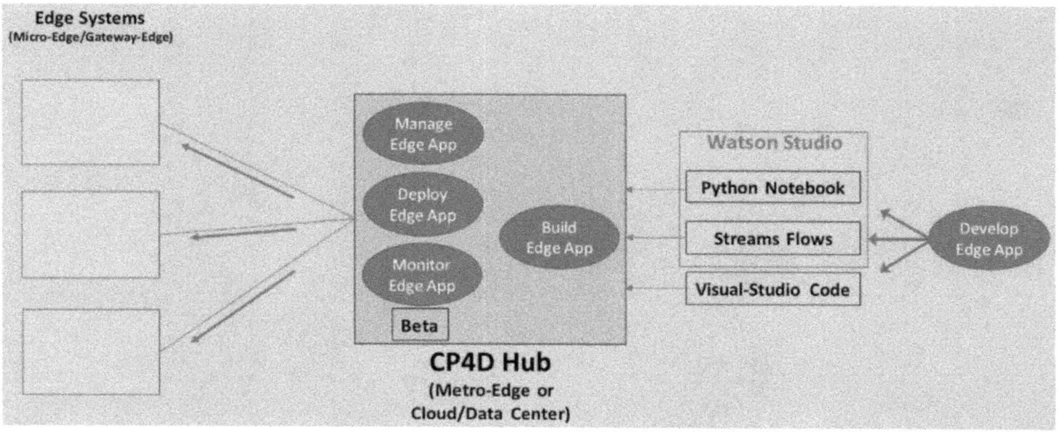

Figure 5.23 – Pushing compute to the edge

We have covered all the objectives of the chapter here. Now, let's look at everything we learned in this chapter by moving on to the *Summary* section next.

Summary

We started this chapter by describing the different types of analytics, understanding the relationship between AI, ML, and DL. We then looked at the current process to source data for analytics, as well as improvements that ModelOps brings to this process.

Having a good understanding of the needs of a diverse data science team, we then went through the different steps in the data science process, using the capabilities/services available as part of the Cloud Pak for Data platform.

In the next chapter, we will explore the platform services catalog and the different types of services—both IBM and third-party capabilities—that help form a rich ecosystem of choices for our customers.

6
Multi-Cloud Strategy and Cloud Satellite

Hybrid multi-cloud support is the foundational pillar of Cloud Pak for Data, and the feature of deploying and running anywhere (private cloud or any public cloud) has been one of Cloud Pak for Data's key differentiators. In this chapter, we will learn more about the supported public and private clouds, the evolution of Cloud Pak for Data as a Service, and IBM's strategy to support managed services on third-party clouds. We will explore both the business and technical concepts behind IBM's multi-cloud support, including a brief overview of IBM Cloud Satellite. This will allow the customer to define an effective multi-cloud strategy by leveraging IBM technology.

We're going to cover the following main topics:

- IBM's multi-cloud strategy
- Supported deployment options
- Cloud Pak for Data as a Service
- IBM Cloud Satellite
- A data fabric for a multi-cloud future

IBM's multi-cloud strategy

IBM is one of the few technology vendors to have embraced a hybrid multi-cloud strategy from day one, and this is evident from the deployment options that are supported by IBM. Being able to deploy anywhere is a key differentiator for Cloud Pak for Data. While software and system deployment options enjoy customer adoption the most, Cloud Pak for Data *as a Service* is IBM's strategic direction in the long run. The as-a-Service managed edition of Cloud Pak for Data was launched in 2020 on IBM Cloud and is now supported on third-party clouds through Cloud Satellite, which we will cover in detail later in the chapter.

The value proposition of *as-a-Service* is that it allows customers to modernize how they collect, organize, analyze, and infuse AI with no installation, management, or updating required. In essence, you can derive all the benefits of an integrated data and AI platform, namely Cloud Pak for Data, without the overhead of managing the infrastructure or software.

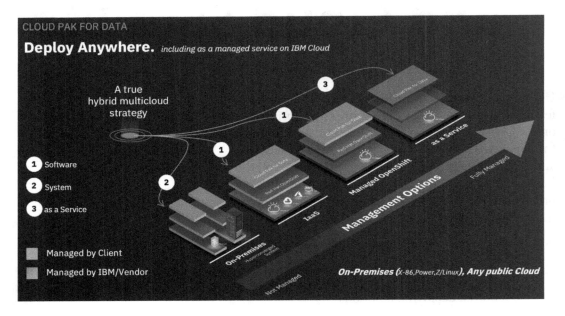

Figure 6.1 – IBM's hybrid multi-cloud strategy

Cloud Pak for Data as a Service is a strategic priority for IBM as there is a growing demand for **Software as a Service (SaaS)**. Furthermore, IBM launched Cloud Satellite to deliver SaaS on-premises and third-party clouds. The objective is to give customers the flexibility of choice so that they can consume software and SaaS on the infrastructure of their choice such as an on-premise private cloud (x86, Power, Z) or a public cloud such as IBM Cloud, AWS, Azure, and so on. Next, we will cover supported deployment options.

Supported deployment options

Cloud Pak for Data comes in three deployment options: software, system, and as-a-Service, offering significant choice to customers as to where they deploy and run their software and how they manage Cloud Pak for Data:

- Cloud Pak for Data **software** runs on any private or public cloud, including AWS, Azure, and GCP. Also supported are managed **OpenShift on IBM Cloud (ROKS)**, managed **OpenShift on AWS (ROSA)**, and managed **OpenShift on Azure (ARO)**.

- Cloud Pak for Data **system** is a true plug-and-play, all-in-one, enterprise data and AI platform that comes with all the necessary hardware and software components.

- Cloud Pak for Data *as a Service* runs on IBM Cloud and third-party clouds such as AWS through **IBM Cloud Satellite**.

For enterprise customers who cannot yet embrace the public cloud or would prefer a non-IBM cloud, the options are as follows:

- **Private cloud**: Cloud Pak for Data software, with its embedded Red Hat OpenShift, enables customers to deploy and operate a cloud-native data and AI platform in their own data center. Alternatively, clients can opt for Cloud Pak for Data system, an integrated appliance with hardware and software pre-installed, optimized for end-to-end data and AI workloads.

- **Third-party clouds**: Cloud Pak for Data software is supported on all major public clouds, including AWS, Azure, and GCP, either directly on their cloud infrastructure or on their managed OpenShift offering, such as **Red Hat OpenShift on AWS (ROSA)** or **Azure Red Hat OpenShift (ARO)**.

Managed OpenShift

Red Hat has launched managed OpenShift offerings on all major cloud providers, including IBM Cloud, AWS, Azure, and GCP. Customers can run their workloads, including Cloud Pak for Data ones, without worrying about the maintenance, administration, and upgrading of OpenShift, which is not so easy for non-technology users.

IBM Cloud Pak for Data is currently available and supported on three major managed OpenShift services. They are IBM Red Hat ROKS, Amazon ROSA, and Microsoft ARO. Managed OpenShift's main value proposition includes four key points:

The main value proposition of this service spans 4 key points:

- **Time saved**: Red Hat takes care of OpenShift so that customers can focus on their data and AI workloads. Furthermore, delivery pipelines are pre-integrated with user-friendly management tools, simplifying the effort required to manage and administer Cloud Pak for Data.

- **Consistency**: Multi-cloud with Red Hat OpenShift works everywhere, including on public clouds and on-premises, and hybrid support enables flexibility of deployment. Also, it allows customers to focus and build skills in one Kubernetes distribution (namely OpenShift) that spans all deployment options.

- **Portability**: Run Red Hat OpenShift workloads consistently with any major cloud provider, offering customers the choice and flexibility they desire. More importantly, this gives customers the ability to migrate workloads from one cloud provider to another if needed.

- **Scalable Red Hat OpenShift**: Control versions and licensing from a single Red Hat account irrespective of where it is deployed.

Here are all the items managed by the provider as part of Managed OpenShift:

- Automated provisioning and configuration of infrastructure (compute, network, and storage).

- Automated *installation and configuration of OpenShift*, including high-availability cross-zone configuration.

- Automatic upgrades of all components (operating system, OpenShift components, and in-cluster services).

- Security patch management for operating system and OpenShift.

- Automatic failure recovery for OpenShift components and worker nodes.

- Automatic scaling of OpenShift configuration.

- Automatic backups of core OpenShift ETCD data.

- Built-in integration with the cloud platform – monitoring, logging, Key Protect, IAM, Activity Tracker, Storage, COS, Security Advisor, Service Catalog, Container Registry, and Vulnerability Advisor.

- Built-in load balancer, VPN, proxy, network edge nodes, private cluster, and VPC capabilities.

- Built-in security, including image signing, image deployment enforcement, and hardware trust.

- 24/7 global **Site Reliability Engineering** (**SRE**) to maintain the health of the environment and help with OpenShift.

- The global SRE team has lots of experience and skill in IBM Cloud infrastructure, Kubernetes, and OpenShift, resulting in much *faster problem resolution.*

- Automatic compliance for your OpenShift environment (HIPAA, PCI, SOC2, and ISO).

- Capacity expansion with a single click.

- Automatic multi-zone deployment in **Multi-Zone Regions** (**MZRs**), including integration with **Cloud Internet Service** (**CIS**) to do cross-zone traffic routing.

- Automatic operating system performance tuning and security hardening.

AWS Quick Start

"Quick Start" on AWS refers to templates (scripts) that automate the deployment of workloads – in this case, Cloud Pak for Data. A Quick Start launches, configures, and runs compute, network, storage, and all other related AWS infrastructure required to deploy Cloud Pak for Data on AWS in 3 hours or less. This is a significant value proposition for customers interested in running Cloud Pak for Data on AWS. It saves time through automation (eliminates many of the steps required for manual installation configuration) and helps implement AWS best practices by design.

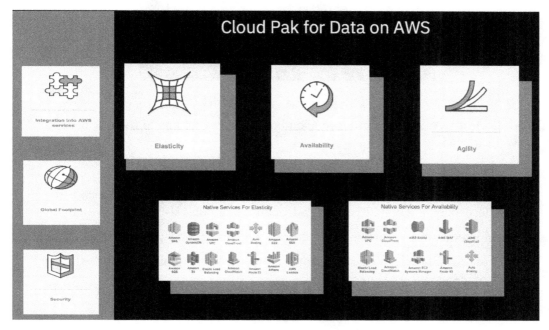

Figure 6.2 – Cloud Pak for Data on AWS – value proposition

Other benefits of AWS Quick Start include a free trial of Cloud Pak for Data for up to 60 days and a detailed deployment guide. However, it requires an AWS account – the customer is responsible for infrastructure costs. For storage, the customer has two options: **OpenShift Container Storage (OCS)**, which was recently renamed **OpenShift Data Foundations (ODF)**, or **Portworx**.

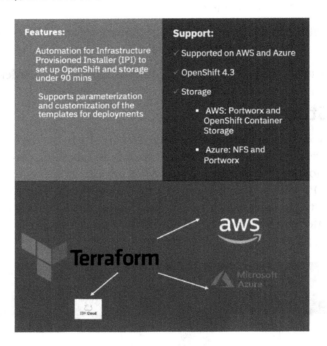

Figure 6.3 – Terraform automation

Also available on AWS, Azure, and IBM Cloud is Terraform automation, whose benefits include the following:

- Easy full-stack deployment to allow consistency and repeatability

- Increased productivity with Infrastructure as Code to write and execute code to define, deploy, and update your infrastructure in minutes

- Idempotency, with code that works correctly no matter how many times you run it

- Self-service through automation for the entire deployment process to schedule and kick off deployments as needed

Azure Marketplace and QuickStart templates

Azure Marketplace is an online applications and services marketplace ideal for IT professionals and cloud developers interested in Cloud Pak for Data software and its services. It enables customers to discover, try, buy, and deploy a solution in just a few clicks. Like AWS, Microsoft requires an Azure account and one of the two supported storage options, namely OCS or Portworx. Also, customers can have a trial at no cost for up to 60 days:

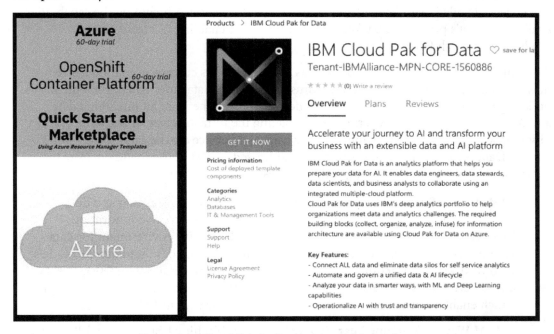

Figure 6.4 – Cloud Pak for Data on Azure Marketplace

Let's move on to cover Cloud Pak for Data as a service.

Cloud Pak for Data as a Service

Most of IBM's data and AI products are currently available as a service and packaged under a single Cloud Pak for Data subscription that is consumption-based, allowing customers to only pay for what they use. Furthermore, to accelerate the modernization of existing workloads and help customers with moving to the cloud/managed services, IBM has launched an initiative called **Hybrid Subscription Advantage (HSA)** that offers existing software customers discounts to use SaaS instead of software. In this section, we will cover a detailed overview of Cloud Pak for Data as a Service, including the capabilities available, how it's priced and packaged, and how IBM is enabling its customers to modernize through HSA.

As mentioned before, Cloud Pak for Data as a Service allows customers to modernize how they collect, organize, analyze, and infuse AI with no installation, management, or updating required. In other words, it helps deliver all the benefits of an integrated data and AI platform without the overhead of managing the infrastructure or software. The different services constituting Cloud Pak for Data as a Service are the same as for Cloud Pak for Data software, with the existing gaps addressed as part of the roadmap. Here is a short list of available services along with their descriptions and value propositions:

1. **IBM Watson Studio**: Watson Studio democratizes machine learning and deep learning to accelerate the infusion of AI in your business to drive innovation. Watson Studio provides a suite of tools and a collaborative environment for data scientists, developers, and domain experts. It includes capabilities to deploy models with IBM Watson Machine Learning and manage them at scale, ensuring accuracy and governance with Watson OpenScale. Last but not least, it comes with Auto AI, which enables normal users to build complex models with just a few clicks.

    a) **IBM Watson Machine Learning**: IBM Watson Machine Learning is a full-service IBM Cloud offering that makes it easy for developers and data scientists to work together to integrate predictive capabilities with their applications. The Machine Learning service is a set of REST APIs that you can call from any programming language to develop applications that make smarter decisions, solve tough problems, and improve user outcomes.

b) **IBM Watson OpenScale**: IBM Watson OpenScale tracks and measures outcomes from AI throughout its life cycle and adapts and governs AI in changing business situations. It also helps with drift explainability and bias detection.

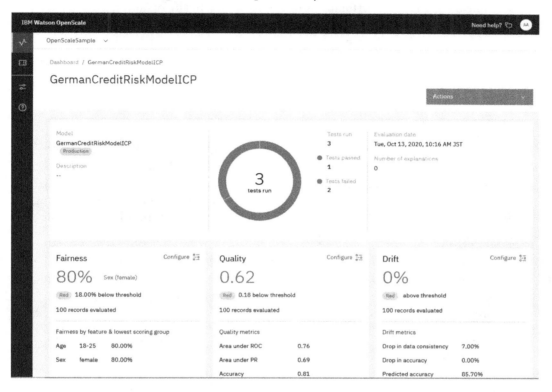

Figure 6.5 – Watson OpenScale

2. **IBM Watson Knowledge Catalog**: Simplify and organize your enterprise data with IBM Watson Knowledge Catalog, making it easy to find, share, and govern data at scale. Customers can create a 360-degree view of their data, no matter where (or in what format) it is stored, auto-discover the data from a myriad of supported data sources, collaborate with fellow users, and control data access by defining policies and monitoring enforcement.

3. **Data Virtualization** enables you to query data across many systems without having to copy and replicate it, saving time and reducing costs. Data Virtualization queries data from its source, simplifying your analytics by providing the latest and most accurate data. The Data Virtualization service automatically organizes your data nodes into a collaborative network for computational efficiency. You can define constellations with large or small data sources. Data is never cached in the cloud or on other devices. Also, credentials for your private databases are encrypted and stored on your local device.

4. **IBM Db2**: A fully managed, highly performant relational data store running the enterprise-class Db2 database engine, IBM Db2 is built to take on the toughest mission-critical workloads on the planet, with advanced features such as adaptive workload management, time travel query, query federation, in-database AI, row/column access control, auditing, and support for JSON, XML, and geospatial datasets. Customers can independently scale and manage the compute and storage requirements for their deployments. Self-service managed backups to object storage with point-in-time recovery allows data to be restored to any specified time while all customer data is encrypted in motion and at rest.

    a) **IBM Db2 Warehouse**: A fully managed elastic cloud data warehouse that delivers independent scaling of storage and compute. It delivers a highly optimized columnar data store, actionable compression, and in-memory processing to supercharge your analytics and machine learning workloads.

5. **IBM DataStage** offers industry-leading batch and real-time data integration to build trusted data pipelines across on-premises and hybrid cloud environments allowing any integration style (ETL or ELT) to prepare data for AI. Please note that ELT stands for **Extract, Transform, Load**, while ELT is extract and load followed by processing. Work with your peers on DataStage Flows and control admin, editor, or viewer access to your projects. Easily perform data integration work in a no-code/low-code environment with a friendly user interface. Scale horizontally or vertically as needed in a secure cloud environment. Take advantage of shared platform connections and integrations with other products in Cloud Pak for Data.

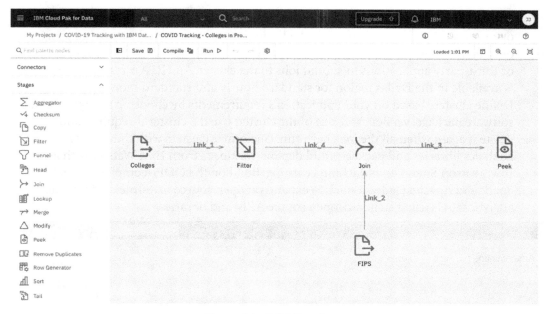

Figure 6.6 – IBM DataStage

6. **IBM Cognos Dashboard Embedded** enables business users and developers to easily build and visualize data with a simple drag and drop interface to quickly find valuable insights and create visualizations on their own.

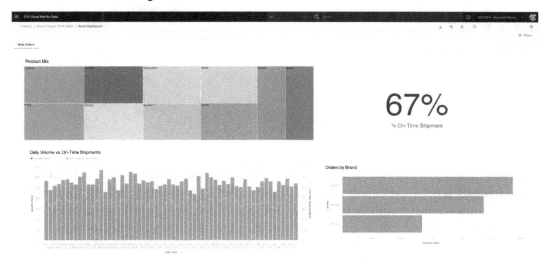

Figure 6.7 – Cognos Dashboard Embedded

7. **IBM Analytics Engine** helps develop and deploy analytics applications using the open source Apache Spark and Apache Hadoop. Customize the cluster using your own analytics libraries and open source packages. Integrate with IBM Watson Studio or third-party applications to submit jobs to the cluster. An HIPAA readiness option is available in the Dallas region for standard hourly and standard monthly plans. Define clusters based on your application's requirements by choosing the appropriate software package, version, and size of the cluster. Use the cluster if required and delete it again when all the jobs have run. Customize clusters with third-party analytics libraries and packages, and deploy workloads from IBM Watson such as IBM Watson Studio and Machine Learning. Build on the ODPi-compliant Apache Spark and Apache Hadoop stack to expand on open source investments. Integrate analytics tools using standard, open source APIs, and libraries.

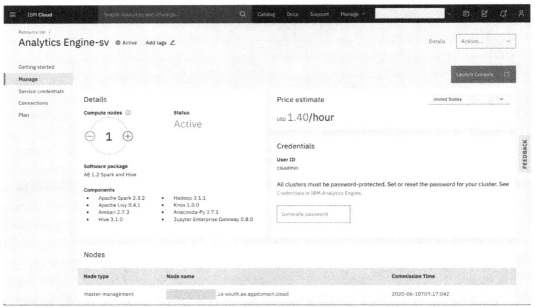

Figure 6.8 – IBM Analytics Engine

8. **IBM Match 360 with Watson** helps quickly build data pipelines for analytics and other data science use cases using master data. Start with your IBM MDM Advanced or Standard Edition entities, or with any data assets containing party information from the knowledge catalog and quickly map and model new attributes to your data model for a more complete view of your customers. The AI-powered matching engine speeds up configuration using statistical methods clients have relied on to produce accurate results. Results can be accessed via RESTful APIs, exported to flat files, or viewed online via the entity explorer. Extend existing master data entities from MDM Advanced or Standard Edition with governed data assets.

Enable text searches and queries against your master data entities. Build service integrations using RESTful APIs or IBM App Connect for use in business applications or data warehouses. Auto-classify and map fields to your data model, and tune your algorithm with data-first, AI-suggested matching attributes. Built-in machine learning and years of experience speed up configuring and tuning party matching algorithms.

9. **Watson Assistant**: Build conversational interfaces into any application, device, or channel. Add a natural language interface to your application to automate interactions with your end users. Common applications include virtual agents and chatbots that can integrate and communicate on any channel or device. Train the Watson Conversation service through an easy-to-use web application designed so you can quickly build natural conversation flows between your apps and users, and deploy scalable, cost-effective solutions.

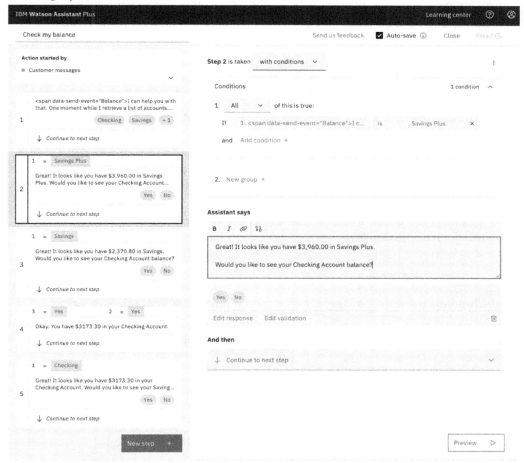

Figure 6.9 – IBM Watson Assistant

a) The **Speech to Text** service converts human voice input into text. The service uses deep learning AI to apply knowledge of grammar, language structure, and the composition of audio and voice signals to accurately transcribe human speech. It can be used in applications such as voice-automated chatbots, analytic tools for customer-service call centers, and multi-media transcription, among many others.

b) The **Text to Speech** service converts written text into natural-sounding speech. The service streams the synthesized audio back with minimal delay. The audio uses the appropriate cadence and intonation for its language and dialect to provide voices that are smooth and natural. The service can be used in applications such as voice-automated chatbots, as well as a variety of voice-driven and screenless applications, such as tools for the disabled or visually impaired, video narration and voice-over, and educational and home-automation solutions.

c) **Natural Language Understanding** helps analyze text and extract metadata from content such as concepts, entities, keywords, categories, sentiment, emotion, relations, and semantic roles. Apply custom annotation models developed using Watson Knowledge Studio to identify industry-/domain-specific entities and relations in unstructured text with Watson NLU.

10. **Watson Discovery** adds a cognitive search and content analytics engine to applications to identify patterns, trends, and actionable insights that drive better decision making. It securely unifies structured and unstructured data with pre-enriched content and uses a simplified query language to eliminate the need for the manual filtering of results. The automated ingestion and integrated natural language processing in the fully managed cloud service remove the complexity of dealing with natural language content. Uncover deep connections in your data by using advanced out-of-the-box AI functions, such as natural language queries, passage retrieval, relevancy training, relationship graphs, and anomaly detection.

11. **Open source databases**:

a) **Cloudant** is a fully managed JSON document database that offers independent serverless scaling of provisioned throughput capacity and storage. Cloudant is compatible with Apache CouchDB and accessible through a simple to use HTTPS API for web, mobile, and IoT applications.

b) **Elasticsearch** combines the power of a full text search engine with the indexing strengths of a JSON document database to create a powerful tool for the rich data analysis of large volumes of data. IBM Cloud Databases for Elasticsearch makes Elasticsearch even better by managing everything for you.

c) **EDB** is a PostgreSQL-based database engine optimized for performance, developer productivity, and compatibility with Oracle. IBM Cloud Databases for EDB is a fully managed offering with 24/7 operations and support.

d) **MongoDB** is a JSON document store with a rich query and aggregation framework. IBM Cloud Databases for MongoDB makes MongoDB even better by managing everything for you.

e) **PostgreSQL** is a powerful, open source object-relational database that is highly customizable. It's a feature-rich enterprise database with JSON support, giving you the best of both the SQL and NoSQL worlds. IBM Cloud Databases for PostgreSQL makes PostgreSQL even better by managing everything for you.

Packaging and pricing

Cloud Pak for Data as a Service is offered as a subscription wherein customers pay for what they use and no more. Also, subscription credits can be used for any of the data and AI services that make up Cloud Pak for Data. This affords a lot of flexibility to customers as they can provision and use all the in-scope services. Each of these individual services is priced competitively to reflect the value they offer and include infrastructure costs such as compute, memory, storage, and networking.

Also, IBM has an attractive initiative called HSA to incentivize existing on-premises customers to modernize to Cloud Pak for Data as a Service. The value in this initiative is that existing software customers will receive a significant discount on Cloud Pak for Data as a Service to account for the value of the software that they have already paid for. This, of course, assumes that customers will stop using other software and instead move to Cloud Pak for Data as a Service.

IBM Cloud Satellite

IBM Cloud Satellite is an extension of IBM Cloud that can run within a client's data center, on an edge server, or on any cloud infrastructure. IBM has been porting all its cloud services to Kubernetes, enabling the different services to function consistently. IBM Cloud Satellite extends and leverages the same underlying concept, running on Red Hat OpenShift as its Kubernetes management environment.

To be more precise, every Cloud Satellite location is an instance of IBM Cloud running on local hardware or any third-party public cloud (such as AWS or Azure). Furthermore, each Cloud Satellite location is connected to the IBM Cloud control plane. This connection back to the IBM Cloud control plane provides audit, packet capture, and visibility to the security team, and a global view of applications and services across all satellite locations. IBM Cloud Satellite Link connects IBM Cloud to its satellite location and offers visibility into all the traffic going back and forth.

IBM's strategy is to bring Cloud Pak for Data managed services to third-party clouds and dedicated on-premises infrastructure using Cloud Satellite. As of 2021, a subset of Cloud Pak for Data managed services, including Jupyter Notebook and DataStage, is available on Cloud Satellite to be deployed on third-party clouds. The packaging and pricing for these services is no different from how it is on IBM Cloud, and included will be the third-party cloud infrastructure to run the services. The ultimate objective here is to deliver a managed service on the infrastructure of your choice.

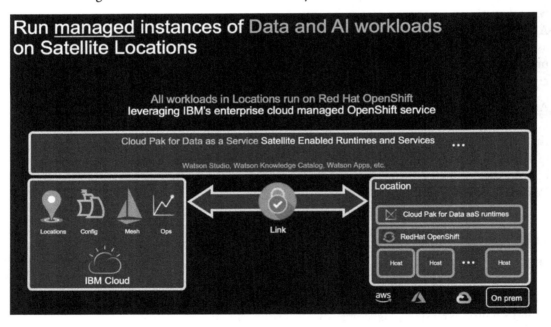

Figure 6.10 – IBM Cloud Satellite

So, how does this all work? Once a Satellite location is established, all workloads in these Satellite locations run on Red Hat OpenShift – specifically IBM Cloud's managed OpenShift service. Services integrated into Cloud Pak for Data as a Service – such as DataStage – can deploy runtimes to these Satellite locations so that these runtimes can be closer to the data or apps these services need to integrate with. A secure connection, called Satellite Link, provides communication between Cloud Pak for Data as a Service and its remote runtimes.

The value proposition of IBM Cloud Satellite specifically for data and AI workloads is threefold:

1. **Data gravity**: Minimize data movement and redundancy, which also helps avoid egress costs. Train AI models closer to where data is located, which removes latency and improves performance.

2. Modernize legacy workloads and realize cloud-native benefits on-premises or on a third-party cloud of your choice:

   Easy to provision and scale up and down.

   Seamless upgrades with negligible downtime.

   Realize the benefits of a managed service.

3. Address regulatory and compliance challenges:

   Share insights without moving data: handle challenges with data sovereignty.

   Comply with GDPR, CCPA, and so on.

Now that we have learned the basics of IBM Cloud Satellite and IBM's approach to multi-cloud, let's look at the **data fabric**.

A data fabric for a multi-cloud future

The cloud is transforming every business, and multi-cloud is the future. To be successful, enterprises will have to access, govern, secure, transform, and manage data across private and public clouds. This is what IBM is addressing using the data fabric, a significant effort that will likely span multiple years. A solid foundation to a data fabric starts with centralized metadata management, data governance, and data privacy, which, when augmented by automation, really helps in amplifying the benefits.

The following image showcases data fabric for a multi-cloud future:

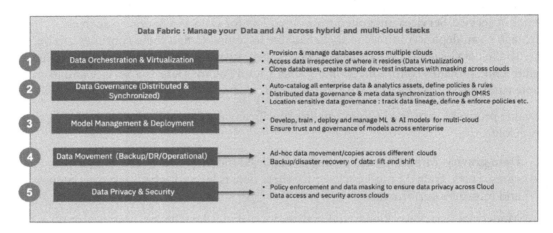

Figure 6.11 – Data fabric for a multi-cloud future

This repeats the last line in the previous paragraph - delete one of them.

Summary

Managed services/SaaS is an absolute must these days, and Cloud Pak for Data has a comprehensive and evolving set of capabilities that addresses end-to-end customer requirements. With simplified packaging and pricing and incentives to modernize, IBM makes it easy for existing and new clients to embrace Cloud Pak for Data as a Service. In this chapter you learned the different deployment options for Cloud Pak for Data, supported third-party clouds and an overview of its managed service (Cloud Pak for Data SaaS offering). You have also learned that IBM's vision is to deliver the Cloud Pak for Data managed service on any infrastructure of your choosing, including third-party cloud providers. This is made possible using Cloud Satellite, IBM's answer to AWS Outposts and Azure Stack/Arc.

Finally, IBM is making significant investments in its data fabric to access, govern, manage, and secure data and AI workloads across clouds to address the evolving requirements of a multi-cloud future.

In the next chapter, we will learn about the Cloud Pak for Data ecosystem, which complements and extends the capabilities included in the base Cloud Pak for Data.

7
IBM and Partner Extension Services

The ecosystem is a critical component of any successful platform, and **Cloud Pak for Data** is no exception. As explained in *Chapter 2, Cloud Pak for Data: A Brief Introduction*, it has a two-tiered packaging structure comprising "a base platform that comes with a set of modular services, but priced and packaged collectively as a single unit, allowing clients to tap into what they need" and "Premium extension services that are optional and priced/ packaged separately."

This chapter focuses on the extension services from **IBM**, also known as cartridges and capabilities from our ecosystem partners. The IBM extension services are tightly integrated into Cloud Pak for Data and include the platform services that enable them to operate independently or as part of an integrated platform. On the other hand, third-party extensions require Cloud Pak for Data and **Red Hat OpenShift**.

In this chapter, we're going to cover the following main topics:

- IBM and third-party extension services
- Collect cartridges: **Db2 Advanced** and **Informix**

- Organize cartridges: Data Stage, Information Server, Master Data Management, and IBM virtual Data Pipeline

- Infuse cartridges: Cognos Analytics, Planning Analytics, Watson Assistant, Discovery, Watson AI for Operations, Watson Financial Crimes Insight, and more

- Modernization upgrades to Cloud Pak for Data cartridges

IBM and third-party extension services

The concept of extension services is best explained using the simple analogy of an iPhone. When someone buys an iPhone, in addition to the hardware and Apple iOS operating system, you also receive several pre-installed and pre-integrated apps from Apple, such as Camera, Maps, Phone, iMessage, Music, and more. These pre-installed and tightly integrated apps enable clients to get started on day one and address the majority of their day-to-day needs. These day-one apps are included as part of the initial purchase. In addition, you can also download third-party premium apps from the App Store that are priced and packaged separately. These are optional apps that address specific requirements and enhance the overall value of an iPhone.

The two-tiered packaging structure of Cloud Pak for Data (shown next) is somewhat analogous to the iPhone. When you buy Cloud Pak for Data, you are entitled to a number of data and AI capabilities required to collect, organize, and analyze data, as well as infuse AI into business processes. Of course, these are modular sets of services, enabling customers to utilize just what they need while retaining the ability to expand or scale them in the future. Also similar to an App Store, Cloud Pak for Data includes an ecosystem of premium services from IBM and third-party vendors enabling customers to optionally tap into a wider set of capabilities. These are called extension services. The IBM extension services are tightly integrated. They include an entitlement to Red Hat OpenShift, and common platform services that enable them to operate independently or as part of a broader platform. As a result, they are also known as cartridges. On the other hand, the third-party services have Cloud Pak for Data and Red Hat OpenShift as prerequisites, and support limited integration with the platform. Nonetheless, they are highly valuable and help complement the capabilities of the base platform:

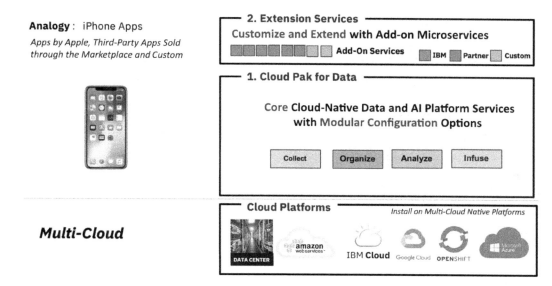

Figure 7.1 – Cloud Pak for Data: two-tier packaging structure

The preceding diagram illustrates the iPhone analogy and how Cloud Pak for Data can be extended and customized with premium extension services from IBM, third-party partners, and custom applications.

Collect extension services

Data is the foundation for businesses to drive smarter decisions and is instrumental in fueling digital transformation. The Collect rung of the AI Ladder is about making data simple and accessible, and Cloud Pak for Data includes a number of base capabilities that address most of the customer requirements, such as data virtualization, Db2 Warehouse, Big SQL, and more. In addition, customers can also tap into the extension services detailed in the following subsections.

Db2 Advanced

IBM Db2 provides advanced data management and analytics capabilities for transactional workloads. Db2 was the world's first relational database and has no processor, memory, or database size limits, which makes it ideal for any size of workload. The Db2 service enables you to create these databases in your Cloud Pak for Data cluster so that you can govern the data and use it for a more in-depth analysis:

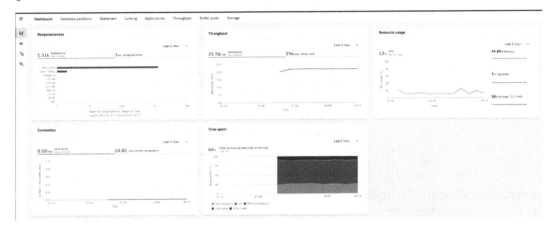

Figure 7.2 – Db2 cartridge

The Db2 database on the Cloud Pak for Data platform can address a number of use case requirements, including the following:

- Persisting and managing transactional data for financial institutions, retail stores, websites, and more

- Replicating transactional data to run analytics without impacting regular business operations

- Ensuring data integrity with an **ACID-compliant database**

- Ensuring low-latency and real-time insight into your business operations

Informix

The **Informix** service provides an Informix database on IBM Cloud Pak for Data. The high-performance Informix engine provides a rich set of features, including **Time Series**, **Spatial**, **NoSQL**, and **SQL** data, which are accessible via **MQTT**, **REST**, and **MongoDB** APIs.

Integrating an Informix database into Cloud Pak for Data can provide the following benefits:

- Providing an operational database for a rapidly changing data model
- Providing lightweight, low-latency analytics integrated into your operational database
- Storing large amounts of data from **Internet of Things (IoT)** devices or sensors
- Storing and serving many different types of content:

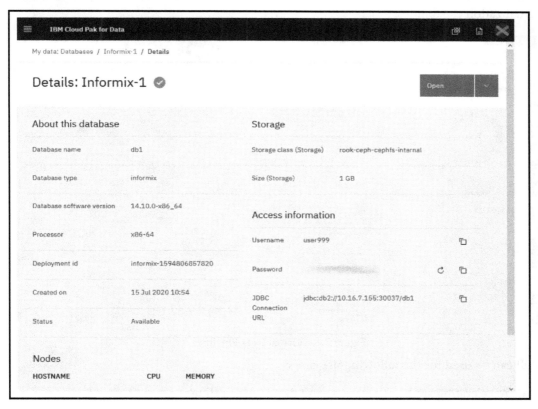

Figure 7.3 – Informix cartridge

Virtual Data Pipeline

Virtual Data Pipeline (**VDP**) enables access to production-like data for data science, analytics, and testing purposes. Traditionally, businesses would take the time-consuming approach of making copies of production data for this purpose, with a higher exposure risk for sensitive data. With the VDP service in Cloud Pak for Data, we are now able to create a single storage-efficient, optionally masked gold copy of a physical data source. This can be further combined with other sources of data and virtualized using **CP4D** data virtualization capabilities:

Figure 7.4 – Virtual Data Pipeline

VDP can be used for the following use cases:

- Accelerating application testing and development time
- Reducing storage costs
- Providing a secure, self-service data provisioning solution for testing, development, and analytics

EDB Postgres Advanced Server

EnterpriseDB Postgres Advanced Server is an integrated, enterprise-ready version of the open source **PostgreSQL** database, with advanced enterprise security, performance diagnostics, Oracle compatibility, and other productivity features:

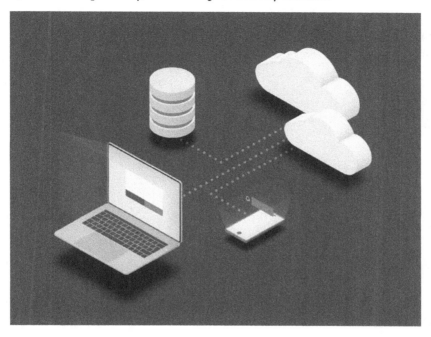

Figure 7.5 – EDB Postgres Advanced Server

EDB Postgres Advanced Server use cases include the following:

- DevOps, schema-less, rapid development with multiple programming language support needed for new applications
- Application modernization that requires multi-modal flexibility and integration with popular data sources
- The need for Oracle compatibility for legacy migrations
- Flexible deployment options enabling a move to the cloud

MongoDB Enterprise Advanced

MongoDB Enterprise Advanced is a highly scalable, document-oriented, NoSQL database that enables dynamic schemas, making it easier to integrate data in certain types of applications. The MongoDB service in Cloud Pak for Data is a highly available database with automatic scaling.

MongoDB Enterprise Advanced also gives you comprehensive operational tooling, advanced analytics and data visualization, platform integrations, and certification:

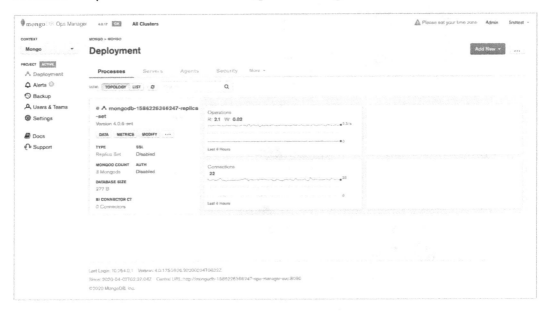

Figure 7.6 – MongoDB Enterprise Advanced

Use cases for MongoDB include the following:

- Legacy application modernization, with a need for evolving schemas
- Making enterprise data available as a service, with a need to combine data from different sources
- Implementing a cloud data strategy, requiring scalability and portability

In the next section, we will look at organizing extension services.

Organize extension services

While the Collect rung of the AI ladder is about putting your data in the appropriate persistence store and enabling easy access, the Organize extension services are all about building a trusted analytics foundation. This entails data prep, data transformation, data governance, data privacy, master data management, and life cycle management of data across the enterprise. While Cloud Pak for Data includes a set of rich capabilities for data prep, data governance, and data privacy as part of the base offering, the rest of the Organize rung is available through the following four extension services (AKA cartridges):

DataStage

Data integration is a critical facet of any enterprise, and it is becoming even more important given the increasing scale and distributed nature of enterprise data assets (often spread across multiple clouds and data centers). IBM **DataStage** is an enterprise-leading **Extract Transform and Load** (**ETL**) offering that has thousands of customers, and is well known for its scalability and operational efficiency. You can use the DataStage service on Cloud Pak for Data to design and run data flows that move and transform data, and this allows you to compose and operationalize your data flow with speed and accuracy. Using an intuitive graphical design interface that lets you connect to a wide range of data sources, you can integrate and transform data, and deliver it to your target system in batches or real time.

Both services provide hundreds of ready-to-use, built-in business operations for your data flows. The high-performance parallel runtime engine underpinning DataStage lets enterprises scale to meet their integration needs despite increasing data volumes and data complexity.

DataStage on Cloud Pak for Data comes in the following two flavors:

- DataStage Enterprise
- DataStage Enterprise Plus

DataStage Enterprise helps design data integration jobs and allows data engineers to transform and move data for multiple sources and targets. You can create reusable job templates, or design data integration flows in the intuitive GUI. Data engineers can also design simple to complex jobs by using a wide range of transformers and stages, while the high-performance runtime engine enables scalability and efficiently integrates large volumes of data.

DataStage Enterprise Plus includes all of the capabilities that DataStage Enterprise has, plus additional useful features for data quality. These features include the following:

- Cleansing data by identifying potential anomalies and metadata discrepancies
- Identifying duplicates by using data matching and the probabilistic matching of data entities between two datasets:

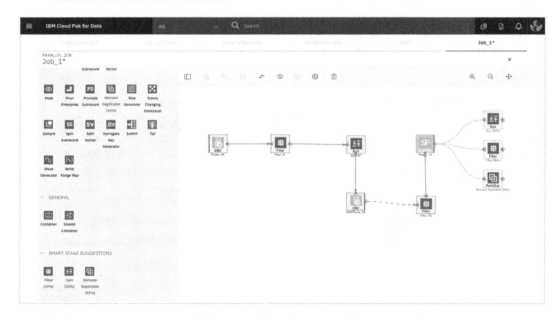

Figure 7.7 – DataStage cartridge

Information Server

The IBM **Information Server** cartridge on Cloud Pak for Data provides IT and **Line-of-Business** (**LOB**) users with a comprehensive suite of capabilities to organize all of your enterprise data. It is a suite of services and includes capabilities to handle data integration, data quality, and data governance. This cartridge is a superset of DataStage and provides seamless integration and a common user experience for the following:

- Automated data discovery
- Policy-driven governance
- Self-service data preparation
- Data quality assessment and cleansing for data in flight and at rest
- Advanced dynamic or batch-oriented data transformation and movement

Master Data Management

Master data is at the heart of every business transaction, application, and decision. The IBM **Master Data Connect** service for IBM Cloud Pak for Data provides a RESTful API so that geographically distributed users can get fast, scalable, and concurrent access to your organization's most trusted master data from IBM **InfoSphere Master Data Management (InfoSphere MDM)**.

InfoSphere MDM acts as a central repository to manage, store, and maintain master data across your organization. InfoSphere MDM provides the following:

- A consolidated, central view of an organization's key business facts

- The ability to manage master data throughout its life cycle:

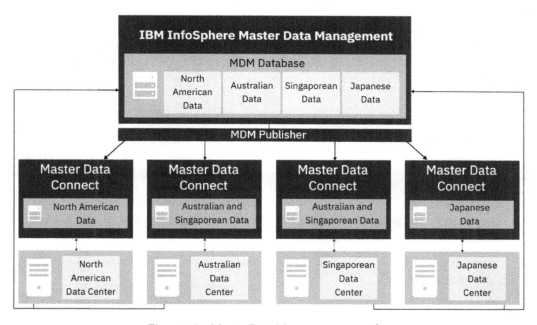

Figure 7.8 – Master Data Management cartridge

InfoSphere MDM version 11.6.0.11 or above supports publishing to Master Data Connect using the IBM **MDM Publisher** capability. With MDM Publisher, you can complete one-time bulk loads of master data into Master Data Connect or set up ongoing data synchronization. When you configure Master Data Connect data subscriptions in the user interface, MDM Publisher creates, runs, and manages the resulting data load jobs.

After your organization's master data is loaded, Cloud Pak for Data users or applications can access the trusted data directly by using the powerful API. This allows them to do the following:

- View master data records.

- Run full text or property query searches on your master data.

- Run graph database queries on your master data.

- View relationships between master data records.

- View the logical model of your data.

Analyze cartridges – IBM Palantir

One of the core use cases of Cloud Pak for Data is data science, and the base platform includes a rich set of capabilities to build, deploy, manage, and govern AI models underpinned by IBM **Watson Studio**. In addition, Cloud Pak for Data includes two premium services (AKA cartridges) to complement the core capabilities.

IBM Palantir

The new IBM **Palantir** cartridge for Cloud Pak for Data will streamline and accelerate how organizations infuse predictive, data-driven insights into real-world decision making and business operations, across a variety of industry verticals. The IBM Palantir service helps with the following:

- Delivering AI-for-business capabilities in an easy-to-deploy-and-manage "no-code, low-code" environment

- Simplifying how data and AI models are connected to business decisions, with trust and explainability

- Informing better decisions by automatically mapping data to business meaning and industry context

- Automating how data is collected, organized, and analyzed across hybrid cloud and data landscapes

- Consuming services from a broad ecosystem of data and AI capabilities, including IBM Watson:

Figure 7.9 – IBM Palantir for Cloud Pak for Data

In the next section, we will review the cartridges that make up the Infuse rung of the AI Ladder.

Infuse cartridges

The Infuse rung of the AI Ladder is about business outcomes and being able to apply AI and data-driven insights to decision making at scale. It entails leveraging AI to automate and optimize a wide range of business use cases in real time, to help advance business objectives. The Infuse rung consists of seven cartridges that help address a variety of business use cases. These cartridges are discussed in the following subsections.

Cognos Analytics

Self-service analytics, infused with AI and machine learning, enable you to create stunning visualizations and share your findings through dashboards and reports.

The **Cognos Analytics** service makes it easier for you to extract meaning from your data, with features such as the following:

- Automated data preparation
- Automated modeling
- Automated creation of visualizations and dashboards
- Data exploration

The service also enables you to easily share your findings with others through stunning visualizations, dashboards, reports, and interactive stories:

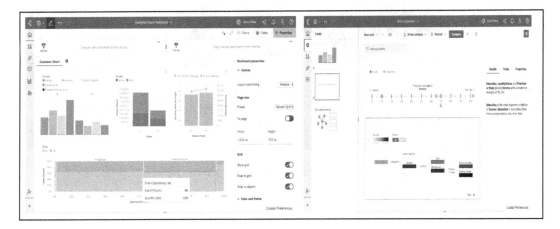

Figure 7.10 – Cognos Analytics cartridge

Planning Analytics

Easily create more accurate plans, budgets, and forecasts using data from across your business. A good plan starts with good data. Ensure that your plans are based on data from across your business with IBM **Planning Analytics**, powered by **TM1®**. Planning Analytics is an AI-infused solution that pulls data from multiple sources and automates the creation of plans, budgets, and forecasts. Planning Analytics integrates with **Microsoft Excel** so that you can continue to use a familiar interface while moving beyond the traditional limits of a spreadsheet.

Infuse your spreadsheets with more analytical power to build sophisticated, multi-dimensional models that help you to create more reliable plans and forecasts. The Planning Analytics service includes the following:

- Easy-to-use visualization tools

- Built-in data analytics and reporting capabilities

- What-if scenario modeling that helps you understand the impact of your decisions:

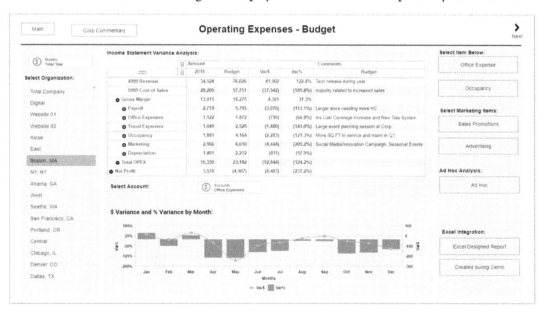

Figure 7.11 – IBM Planning Analytics cartridge

Watson Assistant

An AI assistant accesses and "learns from" hundreds of customer touchpoints while sharing this data with other systems in an integrated way, so that the customer enjoys a consistent experience and is made to feel that the company *knows* them. Well-executed AI-powered chatbots improve the **Net Promoter Score (NPS)** through speed-to-resolution and round-the-clock accessibility, and through the channels in which users are most comfortable, be it web chat, text messages, social media sites, or a phone call to customer services. The **Watson Assistant** service enables you to go beyond chatbots to build conversational interfaces into your apps, devices, and channels. Most chatbots mimic human interactions, but when there is a misunderstanding, users become frustrated. Watson Assistant does more for your users. It knows when to search for an answer in a knowledge base, when to ask for clarification, and when to direct users to an actual person. Watson Assistant can run on any cloud, so you can bring AI to your data and apps, wherever they live:

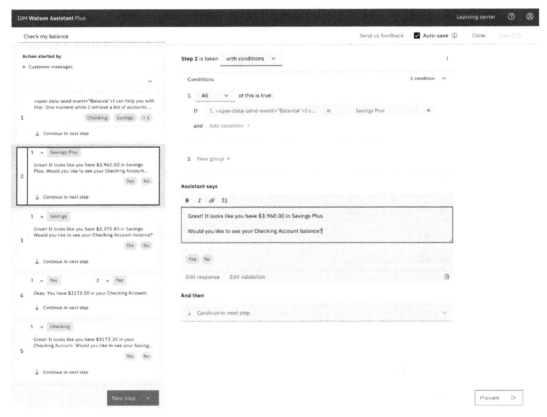

Figure 7.12 – IBM Watson Assistant cartridge

Watson Assistant leverages and includes Watson Discovery for some of its capabilities.

Watson Discovery

The IBM **Watson Discovery** service is an AI-powered search and content analytics engine that enables you to find answers and uncover insights that hide in your complex business content. With the Smart Document Understanding training interface, Watson Discovery can learn where answers live in your content based on a visual understanding of your documents. Want to do more? The following Watson services can help:

- Enhance Watson Discovery's ability to understand domain-specific language with Watson Knowledge Studio.

- Showcase your answers to users through conversational dialog driven by Watson Assistant:

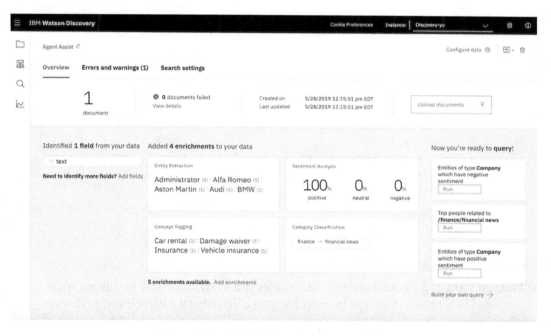

Figure 7.13 – IBM Watson Assistant cartridge

Watson API Kit

The **Watson API Kit** cartridge on Cloud Pak for Data includes the services shown in the following subsections.

Watson Knowledge Studio

The **Watson Knowledge Studio** service uses AI to identify entities, relationships, and other linguistic elements that are unique to your industry. In essence, it helps to teach Watson the language of your domain:

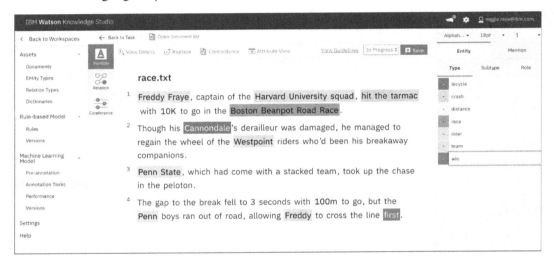

Figure 7.14 – IBM Watson Knowledge Studio

Watson Language Translator

The **Watson Language Translator** service enables you to translate text into more than 45 languages. In addition, with the Watson Language Translator API, you can perform the following:

- Identify the language used in a document.

- Translate documents, such as Word documents and PDF files, into other languages with the same format and layout as the original document (this feature is currently in beta):

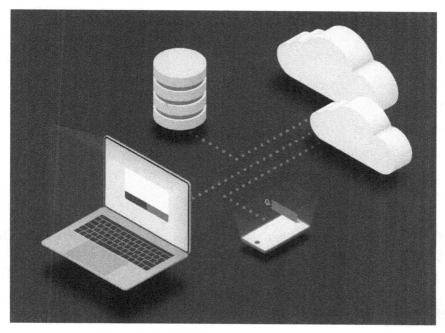

Figure 7.15 – IBM Watson Knowledge Translator

In a nutshell, Watson Language Translator helps you make your application or chatbot accessible to a global audience.

Watson speech services

The **Watson Speech to Text** service provides an API that enables you to add speech transcription services to your applications. The service uses information about language structure and audio signals to create transcriptions. The service can also perform the following:

- Identify acoustically similar alternative words.

- Provide transcription confidence levels.

- Include the audio timestamp for each word in the transcription.

- Enable you to redact PCI data for additional security.

The **Watson Text to Speech** service provides an API that converts written text to natural-sounding speech in a variety of languages and voices. For example, you can use the service to do the following:

- Improve customer experience and engagement by interacting with users in multiple languages and tones.
- Increase content accessibility for users with different abilities.
- Provide audio options to avoid distracted driving.
- Automate customer service interactions to increase efficiency.

IBM OpenPages

The IBM **OpenPages** cartridge on Cloud Pak for Data is an integrated **Governance, Risk, and Compliance** (**GRC**) suite that empowers managers to identify, manage, monitor, and report on risk and compliance initiatives that span your enterprise.

The OpenPages service provides a powerful, scalable, and dynamic toolset that enables you to tackle the following use cases:

- Business continuity management
- Financial controls management
- Internal audit management
- IT governance
- Model risk governance
- Operation risk management
- Policy management
- Regulatory compliance management
- Third-party risk management

The following screenshot showcases the IBM OpenPages service with Watson:

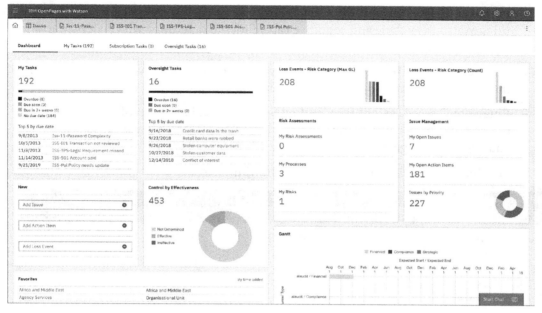

Figure 7.16 – IBM OpenPages for Cloud Pak for Data

Financial Crimes Insight

IBM **Financial Crimes Insight** can be used to simplify the process of detecting and mitigating financial crimes with AI and regulatory expertise. Financial Crimes Insight combines AI, big data, and automation with input from regulatory experts to make it easier to detect and mitigate financial crimes. Install the base offering, Financial Crimes Insight, to proactively detect, intercept, and prevent attempted fraud and financial crimes.

Then, install one or more of the following optional packages depending on your use case:

- **Financial Crimes Insight for Alert Triage**: Enable analysts to quickly assess alerts using Watson's analytic and cognitive capabilities to determine which alerts warrant further investigation.

- **Financial Crimes Insight for Claims Fraud**: Uncover suspicious behavior early in the insurance claims process before fraudulent claims are paid.

- **Financial Crimes Insight for Conduct Surveillance**: Identify, profile, and prioritize employee misconduct using holistic surveillance, behavioral analysis, and cognitive insights:

Let's look at IBM Financial Crimes Insight with Watson:

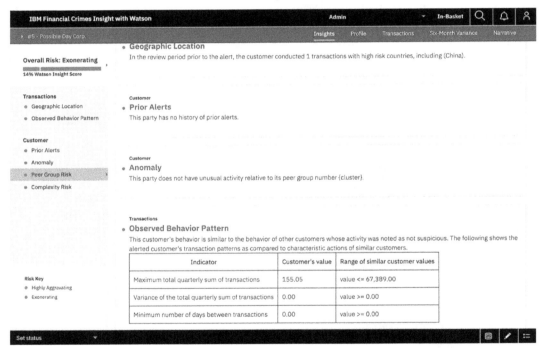

Figure 7.17 – IBM Financial Crimes Insight with Watson

Next, we have IBM Surveillance Insight:

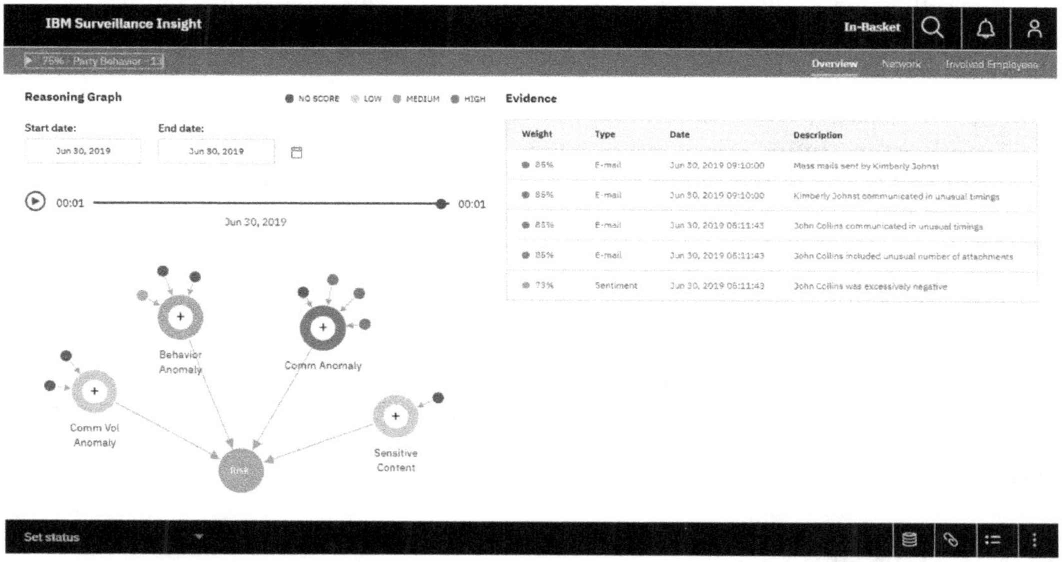

Figure 7.18 – IBM Surveillance Insight

Now that we have covered the Infuse rung and its seven cartridges that help address a variety of business use cases, we'll cover the modernization upgrades to Cloud Pak for Data cartridges.

Modernization upgrades to Cloud Pak for Data cartridges

IBM has thousands of customers running its industry-leading data and AI offerings such as **Db2**, **DataStage**, **Master Data Management**, **SPSS Modeler**, **Decision Optimization**, **Cognos Analytics**, and **Planning Analytics**. Historically, these offerings have had their own independent architecture and ecosystems running on bare metal servers or virtual machines. With the advent of Cloud Pak for Data, all of these offerings are containerized and available as a service on Cloud Pak for Data. Modernizing a standalone offering to a cartridge on Cloud Pak for Data unlocks a number of benefits to clients, such as the following:

- **Cloud-native benefits**: The modern cloud-native architecture of Cloud Pak for Data means it's easier to provision and scale a service, seamless upgrades, and administration. This unlocks major cost savings for enterprises.

- **Deployment flexibility**: In addition to the preceding cloud-native benefits, clients also gain the flexibility to deploy on any public or private cloud, which means enterprises are not locked into a given vendor. This deployment flexibility improves further as more enterprises embrace multiple clouds and a hybrid approach to data and AI workloads.

- **Seamless integration**: The end-to-end integration of services and the tightly orchestrated data fabric ensure self-service, ease of use, and automation, all of which can pay significant dividends to the enterprise.

IBM has recently launched an initiative to modernize existing offerings to Cloud Pak for Data cartridges by paying a nominal trade-up fee. Customers will also receive some incremental entitlement to Cloud Pak for Database licenses. This gives customers a straightforward path to upgrade while keeping costs to a bare minimum. The increased value to clients by modernizing their existing standalone offerings to Cloud Pak for Data services also includes an option to convert standalone licenses to Cloud Pak for Data cartridge licenses at a pre-agreed price of conversion.

Extension services

In addition to the base capabilities and cartridges with Cloud Pak for Data, we also have a thriving ecosystem of partners with whom IBM has co-sell and re-sell relationships. While co-sell partners are tactical and engaged on a deal-by-deal basis, re-sell partners are more strategic with IBM part #s that can be sold by IBM sellers. Given the extensive ecosystem, we will only cover the re-sell partners in detail in the next subsections.

Cambridge Semantics

AnzoGraph DB is a native, massively parallel processing **Graph OLAP (GOLAP)** database designed to accelerate data integration and scalable analytics on graph data at performance levels that compare favorably with industry-leading traditional data warehouse analytics products.

CockroachDB

CockroachDB provides a cloud-native, distributed SQL for cloud applications.

Hazelcast

Hazelcast is an in-memory computing platform, consisting of a Hazelcast in-memory data grid and a Hazelcast in-memory stream processing engine. This delivers a scalable, high-throughput, ultra-low-latency solution, purpose-built to process the most mission-critical, next-generation data and AI/ML workloads.

SingleStore

SingleStore is a high-performance distributed, relational, and converged database that supports both transactional and analytical workloads. The database can ingest millions of events per day with ACID transactions while simultaneously analyzing billions of rows of data in relational SQL, JSON, geospatial, and fill-text search formats.

Datameer

Datameer is a data preparation, enrichment, and exploration platform that simplifies and accelerates the time-consuming process of turning complex, multi-source data into valuable, business-ready information. It provides easy point-and-click tools that can accommodate the skillsets of any person who needs to prepare and explore data, or create datasets for analytics, AI, and ML.

Senzing

Senzing is the first plug-and-play, real-time AI for entity resolution and the most advanced entity resolution software available. You can use Senzing to enhance and transform your big data analytics, fraud operations, insider threat, marketing intelligence, risk mitigation, and other operations. Many organizations struggle with a tsunami of big data, and desperately need to resolve entities to gain new insights and make better decisions faster. Senzing makes this possible with its next-generation entity resolution.

WAND taxonomies

WAND provides the business vocabulary on top of Cloud Pak for Data so that users have a validated source of industry and business function-specific terminology to turn to when starting a data catalog and analytics initiative.

Appen

Appen (formerly Figure Eight) provides a data labeling platform that utilizes machine learning and human annotation to help make unstructured data in text, audio, image, and video formats usable for data science, machine learning, and product teams.

Findability Sciences

The **FP-Predict+** add-on to CPD provides end users with a prediction and forecasting capability that requires little to no manual intervention, no programming, and no manual modeling. With the combination of Cloud Pak for Data components, the solution can work across industries and use cases.

Lightbend

Lightbend provides cloud-native development of at-scale, open source-based streaming systems.

Personetics

Personetics provides data-driven personalization and customer enablement for financial services.

Portworx

Portworx is the market-leading Kubernetes storage platform and provides scalable, performant container storage, backup and disaster recovery, multi-cloud operations, data security, capacity management, and compliance and governance for any data service running on Cloud Pak for Data. This ensures that customers have a flawless operational experience running the software purchased from IBM.

Trilio

TrilioVault provides native backup/recovery for the entire Cloud suite. IT administrators require Trilio's backup/recovery capability as OpenShift is deployed in production, and developers require Trilio's Test/Dev functionality in order to accelerate application agility.

Ecosystem partners with co-sell agreements

The following list provides the names of ecosystem partners with co-sell agreements:

- **BCC Group**: ONE platform for real-time streaming market data
- **Crunchy PostgreSQL**: Open source PostgreSQL distribution for the enterprise
- **Equifax**: Dataset for financial capacity, demographic, and credit information of consumers
- **People Data Labs**: Dataset consisting of B2B and B2C data on person profiles
- **Precisely**: Comprehensive list of all known addresses in the United States and Canada
- **ZoomInfo**: ML-enhanced B2B company data to improve decision making
- **ContractPodAI**: SaaS-based **Contract Lifecycle Management (CLM)**
- **Duality Secureplus Query**: Securely query datasets controlled by third parties
- **DXC Open Health Connect**: Digital health platform for healthcare providers, payers, and the life science industry

- **Equinix**: Enables the running of AI models at the point of data ingest to reduce the cost and risk of data transfer

- **Listec**: Solution to analyze and visualize operational SAP data

- **Prolifics**: Predictive prospecting solution

- **Tavant**: Media viewership analytics accelerator

- **Dulles Research**: Plugin for the automatic conversion of SAS code to open source code

- **NetApp**: Hybrid multi-cloud data services that simplify the management of applications and data across cloud and on-premises environments

Summary

Cloud Pak for Data has a vibrant IBM, open source, and third-party ecosystem of services. In this chapter, we covered in depth the capabilities of IBM premium services (also called cartridges), organized across the four rungs of the AI Ladder, namely, Collect, Organize, Analyze, and Infuse.

In the next chapter, you will learn about the importance of the partner ecosystem, the different tiers of the business partners, and how clients can benefit from an open ecosystem on Cloud Pak for Data.

8
Customer Use Cases

Cloud Pak for Data is a data and AI platform with a wide variety of services that span the different steps of the **artificial intelligence (AI)** ladder; that is, Collect, Organize, Analyze, and Infuse. These include services from IBM, open source, and third-party vendors that allow customers the flexibility to pick services based on their corporate preference and use case requirements.

This vibrant ecosystem and the lego block approach to dynamically using different combinations of services in the platform opens up a world of possibilities. The platform also benefits from the cloud-native architecture, wherein individual services can be scaled up or down dynamically to allow for consumption across different use cases.

Every organization has different personas that interact with data and make use of it for a variety of reasons – this often requires manipulating data and combining data across different systems to create new desired datasets. These derived datasets can be very valuable to business consumers in the organization but are often not shared effectively, leading to multiple copies of data and increased risk exposure. Cloud Pak for Data comes with a centralized governance catalog that enables such cross-persona collaboration.

In this chapter, we will explore some of the typical customer challenges related to how Cloud Pak for Data is used to address these challenges, what underlying services and capabilities of Cloud Pak for Data are leveraged, and the role that's played by different user personas as a team to fix the problem statement at hand. These are just some possible patterns; customers have the flexibility to add/combine any of the available services and design variations.

In this chapter, we will be covering the following topics:

- Improving health advocacy program efficiency
- Voice-enabled chatbots
- Risk and control automation
- Enhanced border security
- Unified Data Fabric
- Financial planning and analytics

Improving health advocacy program efficiency

Let's go through the use case by addressing the challenges that customers face in terms of improving the efficiency of their health advocacy program.

The **customer challenge** was for a non-profit health care provider network and health plan company that wanted to improve the efficiency of their health advocacy program. The goal was to identify members who are at risk and reduce the likelihood of members experiencing adverse events. They wanted to predict the likelihood of an **Emergency Room (ER)** visit in the next 9 months. Their challenge was to cater to the diverse data science teams with demands for different tools/technologies, as well as the support staff that maintain these tools/technologies. They also wanted to operationalize their ML models and scale them across all cohorts over the next year(s).

With Cloud Pak for Data, they were able to catalog information assets, including external datasets, with the centralized governance catalog. They were able to establish a governance framework with individual assets assigned to stewards and access policies that had been defined using governance policies/rules. With access to trusted data, they were able to leverage the Watson Studio Jupyter notebooks to train their ML models and deploy them as Watson Machine Learning deployments. The deployed models were monitored on an ongoing basis for bias and model drift.

The following diagram showcases the ModelOps component:

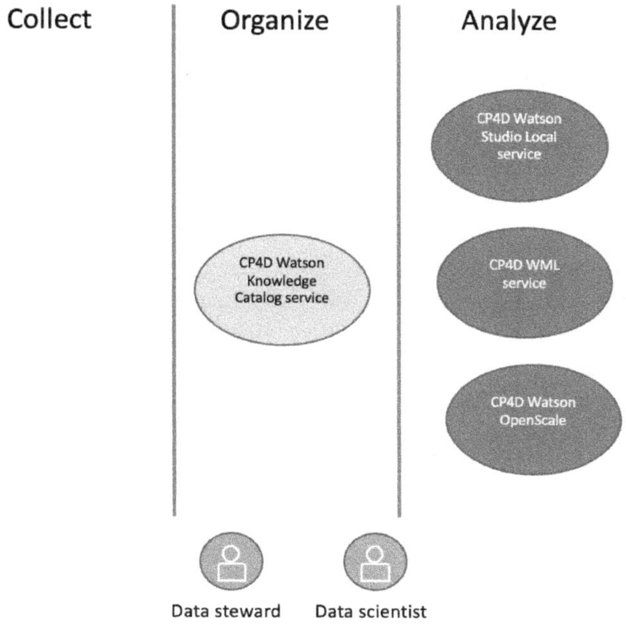

Figure 8.1 – ModelOps component diagram

The featured **Cloud Pak for Data** services include Watson Knowledge Catalog, Watson Studio, and Watson Machine Learning:

- **Data steward**(s) organize data in the catalog and create governance policies and rules. Additionally, they create data rules that obfuscate parts of the datasets to ensure only authorized personnel have access to data.

- **Data scientist**(s) leverage the wide variety of tools/technologies that are available (Jupyter notebooks, **SPSS modeler**, **RStudio**, and so on) to prepare data and train AI models.

The outcome is that, based on member risk level (high, medium, or low risk), this customer was able to prioritize outreach activities. Cloud Pak for Data helped save lives by helping to proactively prioritize the relevant patients based on their risk exposure.

Next, we will cover another customer use case related to voice-enabled chatbots.

Voice-enabled chatbots

The customer challenge included a public petroleum and natural gas company that was struggling to locate consistent and complete information on the services provided by the **Information Technology (IT)** and Engineering departments. Their existing system was a text-based transactional chatbot lacking governance and adoption due to the quality of responses.

With Cloud Pak for Data, they were able to restructure the text-based transactional chatbot with a voice-enabled, avatar-styled interface for improved user experience and, more importantly, better responses to user queries. This, combined with structured analytics, helped to drastically improve customer satisfaction. They used the SPSS modeler and Jupyter notebooks from the Watson Studio service for structured analytics, Watson Discovery for textual search, the Watson API service for voice transcription, and Watson Assistant for an avatar-styled chatbot experience.

The following diagram represents the AI-powered chatbot component:

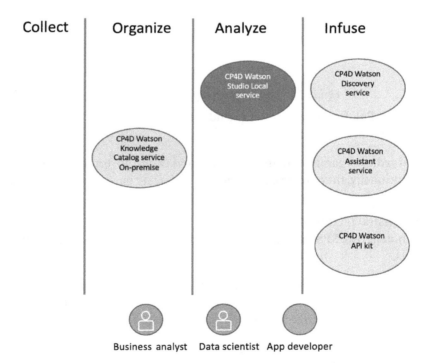

Figure 8.2 – AI-powered chatbot component diagram

The featured **Cloud Pak for Data** services include Watson Knowledge Catalog, Watson Discovery, Watson Assistant, Watson API, and Watson Studio:

- **Data steward**(s) organize data in the catalog and create governance policies and rules. Additionally, they create data rules that obfuscate parts of the datasets to ensure only authorized personnel have access to data.

- **Business analyst**(s) analyze the business needs and define application requirements, including channel-specific interactions.

- **Data scientist**(s) design and train AI models to derive context from existing content and desired dialog(s) for the assistant.

- **Application developer**(s) use trained models to integrate these AI models into applications for the different channels.

The outcome of the use case was that, with this Cloud Pak for Data powered solution, this customer now has an improved service accessibility time. They also helped their employees get the information they needed in real time. Their future plans include delivering multiple AI-powered chatbots while expanding on the overall governance strategy.

Let's move on to the next use case, which is related to risk and control automation.

Risk and control automation

The customer challenge is that a major financial institution requires its audit departments to track and evaluate controls that help them determine their level of risk exposure. Considering that any major bank has 250,000+ controls that are mapped to risks, this was not an easy task.

A good business control clearly states who, what, when, how, and where the control is to be used. Banks often suffer due to poorly defined controls. With Cloud Pak for Data, this customer was able to train an ML model to predict the *Control* quality level of each written control. They used the Watson Studio service to train the model and Watson Machine Learning to deploy and operationalize the model.

The following diagram illustrates the risk and control automation component:

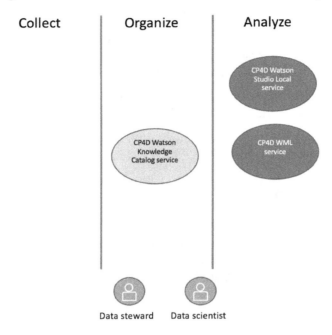

Figure 8.3 – Risk and control automation component diagram

The featured **Cloud Pak for Data** services include Watson Knowledge Catalog, Watson Studio, and Watson Machine Learning:

- **Data steward**(s) organize data in the catalog and create governance policies and rules. Additionally, they create data rules that obfuscate parts of the datasets to ensure only authorized personnel have access to data.

- **Data scientist**(s) design and train ML models and operationalize them for consumption by applications.

The outcome of the use case was that the customer was able to build and deploy an ML model to analyze control definitions and score them based on five factors (who, what, when, how, and why) in a matter of weeks, thus gaining a competitive advantage over their peers.

Our next customer use case is based on enhanced border security.

Enhanced border security

The customer challenge was that a sophisticated customs department, managing how goods were imported and exported into the country, wanted to leverage AI models to identify risky goods entering the country, while also reducing the number of false positives in identifying these risky goods. The client was using a rules-based engine to identify risky goods. However, the vast majority were false positives, which resulted in needless inspections and slowed down the entire customs process.

With Cloud Pak for Data, this customer was able to build a unified platform to handle all of their data science needs – from discovery to deployment. They used platform connectivity to gain access to existing sources of data, the Watson Studio service to train ML/AI models, the Watson Machine Learning service to deploy and operationalize the models, and Watson OpenScale to monitor these models and understand the logic behind them.

The following diagram represents an enhanced border security component:

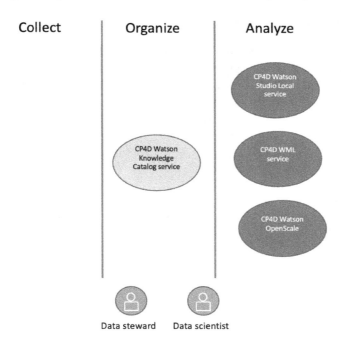

Figure 8.4 – Enhanced border security component diagram

The featured **Cloud Pak for Data** services include Watson Knowledge Catalog, Watson Studio, Watson Machine Learning, and Watson OpenScale:

- **Data steward**(s) organize data in the catalog and create governance policies and rules. Additionally, they create data rules that obfuscate parts of the datasets to ensure only authorized personnel have access to data.

- **Data scientist**(s) leverage the wide variety of tools/technologies (Jupyter notebooks, SPSS modeler, RStudio, and so on) to prepare data, and then train and monitor AI models.

The outcome of the use case was that this customer was able to develop AI models to identify risky goods entering the country. More importantly, they now understand *WHY* something was flagged as risky using the ML model monitoring capabilities from the **Watson OpenScale** service, thus reducing the overall effort and cost to operate the customs agency.

Unified Data Fabric

The customer challenge for this use case was that a leading cancer treatment and research institution wanted to modernize its approach to data with a digital transformation. Despite their exceptional track record regarding patient care, innovative research, and outstanding educational programs, many data sources were still hard to find and consume. There was also a need to archive less frequently used data with appropriate governance controls. They wanted to create an easy-to-use single authoritative platform for their clinical, research, and operational data that would provide self-service data access in a trusted and governed manner.

Unified Data Fabric with **CP4D** allowed them to define centralized governance policies and rules, as well as automated enforcement of data privacy, to ensure only authorized consumers had access to data. The following diagram illustrates the services used to define Unified Data Fabric:

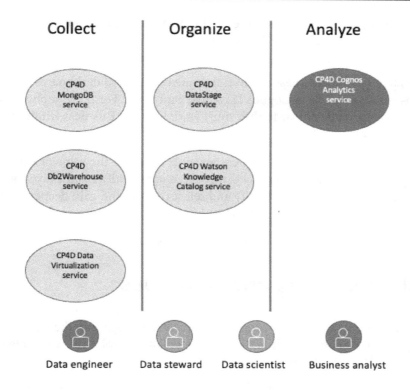

Figure 8.5 – Data Fabric component diagram

The featured **Cloud Pak for Data** services include Db2 Warehouse, MongoDB, data virtualization, Watson Knowledge Catalog, DataStage, and Cognos Analytics:

- **Data engineer**(s) provision data management capabilities and populate data – they use a MongoDB instance for operational data, a Db2 Warehouse instance for storing analytic data, and a data virtualization instance to virtualize data residing in the CP4D platform and other sources of data across the enterprise.

- **Data engineer**(s) provision **Extract**, **Transform**, and **Load** (**ETL**) capabilities with DataStage. This allows data to be sourced from a variety of sources, transforming it with a choice of operators and then loading it to the target repositories.

- **Data steward**(s) organize data in the catalog and create governance policies and rules. Additionally, they create data rules that obfuscate parts of the datasets to ensure only authorized personnel have access to data.

- **Data analyst**(s) and **Business analyst**(s) create reports and dashboards using the easy-to-use drag and drop Cognos Dashboard interface. For more sophisticated reports/scorecards/dashboards, they must work with the **Business Intelligence** (**BI**) team and leverage the Cognos Analytics BI service.

The outcome of this use case is that the client was able to realize the value of Unified Data Fabric for improving the ease of data access through a governed semantic layer on Cloud Pak for Data. Their ability to easily identify and select cohorts and gain deeper insights were greatly improved, and the platform allowed for more self-service analytics.

Next, we have a customer use case on financial planning and analytics.

Financial planning and analytics

The customer challenge was that an online distribution company wanted to scale its planning operations and bring cross-organizational collaboration into the planning process. They had challenges in scaling their existing planning solution, which also lacked effective collaboration capabilities.

Traditionally, all businesses do **Financial planning and analytics** (**FP&A**). This involves multiple finance teams using hundreds of interconnected spreadsheets for planning and reporting. This is a very time-consuming, error-prone process and limits the amount and number of scenarios that can be used. In this era, organizations are looking to go beyond finance to a more collaborative and integrated planning approach known as *FP&A*. This involves multiple teams throughout the organization all collaborating with the goal of gaining forecast accuracy and the ability to quickly respond to changes in market conditions.

The planning analytic service in CP4D helps us move past this manual planning, budgeting, and forecasting process to an automated AI-based process. You can identify meaningful patterns and build what-if scenarios to create models. This, combined with the integrated Cognos Analytics service, enables self-service analytics with stunning visualizations and reports. CP4D, with its data management capabilities and centralized governance catalog, provides access to trusted data for analytics.

The following diagram illustrates the services that are used to deliver *FP&A*:

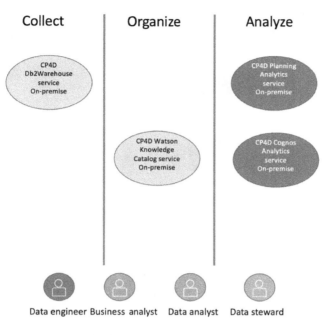

Figure 8.6 – AI for financial operations component diagram

The featured **CP4D** services include Db2 Warehouse, Watson Knowledge Catalog, Planning Analytics, and Cognos Analytics:

- **Data engineer**(s) provision data management capabilities and populate data.

- **Data steward**(s) organize data in the catalog and create governance policies and rules. Additionally, they create data rules that obfuscate parts of the datasets to ensure only authorized personnel have access to data.

- **Data analyst**(s) and **Business analyst**(s) define cubes for planning and also create reports and dashboards using the Cognos Analytics BI service.

Summary

Cloud Pak for Data is a data and AI platform with a wide variety of services that span data management, data governance, and analytics. This gives us options for defining several different use cases using different combinations of services. In this chapter, we looked at some examples of customer use cases that addressed the different challenges of customers and the data and AI platform. In the next chapter, we will provide a technical overview of the platform, including how to manage and administrate it.

Section 3: Technical Details

This section of the book focuses on the technical architecture of Cloud Pak for Data, with emphasis on security, governance, HA/DR, multi-tenancy support, administration, and troubleshooting tips, along with some deployment best practices.

> **Note:**
> The next 3 chapters are highly technical in nature. Hence, for the benefit of the readers who are new to the forthcoming topics, we have included additional references and citations to the end of these chapters, both as introductory material and for advanced insight into these topics, left to the discretion of the reader to pursue (as needed).

This section comprises the following chapters:

- *Chapter 9, Technical Overview, Management, and Administration*
- *Chapter 10, Security and Compliance*
- *Chapter 11, Storage*
- *Chapter 12, Multi-Tenancy*

9
Technical Overview, Management, and Administration

Cloud Pak for Data is a data and AI platform that enables enterprises to tap into their data and accelerate the adoption of AI to drive better business outcomes. This platform helps them modernize their data access mechanisms, organize data for trusted analytics, develop machine learning models, and operationalize AI into their day-to-day business processes. This chapter provides an insight into the foundational aspects of Cloud Pak for Data, the technology stack that powers it, and how services integrate to deliver the key capabilities needed to be successful with AI in the enterprise.

The platform serves several user personas, enabling them to collaborate easily with each other, and helps break down silos within the enterprise. It does this by abstracting the infrastructural elements and by promoting the integration of services to provide a seamless experience for all end users. Enterprises do not start off by using all services in the platform, but rather focus on specific use cases and then gradually expand to other scenarios as needed. You can have different groups of users working on different parts of a solution on Cloud Pak for Data while collaborating and sharing assets as needed. Enterprises may even have entire departments working on different use cases, with built-in isolation, and yet provide for appropriate quality of service to each such *tenant*. Hence, the platform also enables a plug-n-play mechanism and multi-tenant Deployment models that support the extension of the same foundational platform with additional IBM and/or third-party services.

In this chapter, we will be exploring these different aspects:

- Architecture overview
- Infrastructure requirements, storage, and networking
- Foundational services and the control plane
- Multi-tenancy, resource management, and security
- Day 2 operations

Technical requirements

For a deeper understanding of the topics described in this chapter, the reader is expected to have some familiarity with the following technologies and concepts:

- The Linux operating system and its security primitives
- Cloud-native approaches to developing software services
- Virtualization, containerization, and Docker technologies
- Kubernetes (Red Hat OpenShift Container Platform is highly recommended)
- Storage provisioning and filesystems
- Security hardening of web applications, Linux hosts, and containers
- The Kubernetes operator pattern (highly recommended)

- The **kubectl and oc** command-line utilities, as well as the YAML and JSON formats needed to work with Kubernetes objects

- An existing Cloud Pak for Data installation on an OpenShift Kubernetes cluster, with administration authority

The sections will also provide links to key external reference material, in context, to help you understand a specific concept in greater detail or to gain some fundamental background on the topic.

Architecture overview

In *Chapter 1, The AI Ladder: IBM's Prescriptive Approach*, and *Chapter 2, Cloud Pak for Data – Brief Introduction*, you would have seen why we needed such a platform in the first place. In this section, we will start by exploring what exactly we mean by a data and AI platform in the first place. We will outline the requirements of such a platform and then drill down further into the technical stack that powers Cloud Pak for Data.

Characteristics of the platform

Enterprises need a reliable and modern platform founded on strong cloud-native principles to enable the modernization of their businesses. Let's look at what is expected from such a modern platform in general terms:

- **Resiliency**: Continuous availability is a key operational requirement in enterprises. Users need to be able to depend on a system that can self-heal and failover seamlessly without needing a system administrator to manually intervene.

- **Elasticity**: Enterprises expect the platform to grow along with their use cases and customers. Hence, a scalable platform is yet another key requirement. We also need a platform that can support workload *bursting* on demand, in other words, grow automatically when new workloads come in and shrink back when they are completed. Workloads could just be an increased concurrency with more users or background, even scheduled, jobs.

- **Cost-effective**: The ability to support the balancing of available compute across different types of workloads and tenants is also important. In general, this also implies that compute can be easily transferred to those workloads that need it at that time; for example, the ability to assign more compute to a nightly data transformation or ML model retraining job while granting more compute to interactive users at daytime. A *production* workload may get a preference compared to a *development* workload. Thus, this requires the platform to control workloads from crossing resource thresholds, ensuring that no noisy neighbors are impacting both performance and reliability.

- **Extensibility**: The platform is expected to help customers start with one use case and expand when needed. This also implies that the platform cannot be monolithic and architecturally support a plug-n-play model where capabilities and services can be turned on as needed. Enabling the integration of various enterprise and third-party systems, as well as leveraging existing investments, is also key for quickly realizing value from the platform.

- **Portability**: Enterprises require a platform that works the same across different Infrastructure-as-a-Service providers, including on-premises. They need to be able to easily burst into the public cloud or across multiple clouds.

- **Integrated**: A data and AI platform needs to be able to serve different user personas with different skill levels and responsibilities. Such a platform needs to facilitate collaborations among end users, empowering them with self-service capabilities that do not require constant IT handholding. This implies a seamless user experience and APIs that abstract the complexity of the implementation from its consumers.

- **Secure**: The platform must enable enterprises to operate securely, enabling end user workloads or services from multiple vendors to function within well-established boundaries. A data and AI platform also needs to ensure that the enterprise is regulatory compliance-ready. Since the economics of infrastructure cost typically implies the use of shared computing and storage resources to support multiple tenants, techniques to ensure data security and isolation of tenant workloads become fundamental to the enterprise.

Technical underpinnings

In earlier chapters, we introduced **Kubernetes** [1] and **Red Hat OpenShift Container Platform** [2] as the core infrastructure powering all Cloud Paks. In this section, we will look at the capabilities of this technology that makes it the ideal vehicle for delivering a data and AI platform.

Kubernetes is essentially a highly resilient and extensible platform for orchestrating and operating containerized workloads. It embodies the microservices architecture pattern, enabling the continuous development and Deployment of smaller, independent services:

- **Resilient**: Kubernetes enables containerized workloads to be clustered on multiple compute nodes. Kubernetes monitors workloads ("Pods") and, via health probes, can auto-restart them when failure is detected. Even when there are compute node failures, it can automatically move workloads to other nodes. This works out well for establishing *multi-zone* availability, where workloads can survive the outage of entire zones.

- **Elasticity**: The scale-out of computing is elegantly handled by introducing additional compute nodes to the cluster. Kubernetes treats the entire set of CPU cores and memory as a pool of resources and can allocate them to Pods as needed. It also supports the concepts of replicas and **load balancing**, where workloads can be scaled out with multiple copies (Pod replicas). Kubernetes can then route requests to different replicas automatically in a round-robin manner, considering those replicas that may not be healthy at any instant.

- **Infrastructure abstraction**: With Kubernetes, workloads are deployed using higher-level constructs and without the need to be aware of the **compute** infrastructure that hosts it – whether on-premises or in the cloud, or whichever hypervisor is in use. Pods are seamlessly scheduled on different worker nodes based on how much compute (CPU cores and memory) is needed. Kubernetes also includes the ability to consume **storage** for persistence using a vendor-independent framework, and this enables enterprises to work with a wider choice of storage technologies, including leveraging existing storage solutions that they may already have in their enterprises. Software-defined **networking**, and the concept of OpenShift routes and ingress, as well as service discovery patterns, also ensure that workloads are well insulated from the complexity of networking, too.

- **Portable and hybrid cloud-ready**: OpenShift Container Platform delivers a standardized Kubernetes operating environment in many cloud Infrastructure-as-a-Service hyper-scaler environments, even offered as a service as well as on-premises on top of both virtual machine and bare metal nodes. Kubernetes thrives as an open source community, and with that comes challenges with different versions and API compatibility. OpenShift ensures a consistent release cycle for Kubernetes with fully validated and *security-hardened* container platforms everywhere. This ensures that enterprises can deploy their applications anywhere they need or move to another cloud, without the danger of being locked down to an initial choice of the Infrastructure-as-a-Service.

- **A rich developer ecosystem**: Kubernetes has quickly become the development platform of choice for cloud-native applications. OpenShift Container Platform further extends that with developer-focused capabilities, including enriching the concept of **continuous delivery/continuous integration (CI/CD)**. With the focus on secure development and operations, OpenShift Container Platform enables rapid innovation at enterprises for developing modern cloud-native apps and solutions.

- **Enterprise and multi-tenancy ready**: Kubernetes, especially with its concept of namespaces, resource quotas, and network policies, can elegantly support isolations between tenants. **OpenShift Container Platform** [2] introduces additional security constructs, including SELINUX host primitives that greatly improve the safe sharing of OpenShift clusters among multiple tenants in a cost-effective manner.

For these reasons and more, IBM Cloud Pak for Data is powered by the Red Hat OpenShift Container Platform. This enables Cloud Pak for Data to provide an open, extensible, reliable, and secure data and AI platform in any cloud.

The operator pattern

The operator pattern [3] is an architectural concept and best practices to build, deliver, and operate applications in Kubernetes. Operators [4] are special *Controllers* deployed in Kubernetes that are meant to manage individual applications in an automated manner. Essentially, operators codify human experiences to manage and correct Kubernetes workloads.

Cloud Pak for Data v4 functionality is delivered as operators [5]. These operators enable users to install and configure *Day1* activities. From a *Day2* perspective, operators also continuously monitor and auto-correct deviations from the norm as well as support the need to scale up/out deployed Cloud Pak for Data services when needed. Operators are also leveraged to upgrade services to newer releases in an automated fashion.

OpenShift Container Platform includes the **Operator Lifecycle Manager (OLM)** [6] functionality that helps the OpenShift Administrator curate operators for various services (IBM, Red Hat, open source, or other third-party vendors) in a standardized manner.

Cloud Pak for Data includes a *Control Plane* layer that enables all services to integrate seamlessly and be managed from a single platform layer. This control plane is deployed using an operator called the Cloud Pak for Data platform operator [5]. A user with appropriate Kubernetes privileges creates a **Custom Resource (CR)** object in a particular Cloud Pak for Data instance namespace to indicate what service should be deployed. In the case of Cloud Pak for Data, a custom resource of the kind *Ibmcpd* causes the Cloud Pak for Data platform operator to trigger the installation of the control plane in that namespace.

Cloud Pak for Data is a fully modular platform and users can install services only when needed. You could even have a single *standalone* product on top of the control plane, but when multiple services are installed, integration happens automatically, enabling end-to-end use cases to be supported. So, the need for a use case drives which services are *provisioned* on top of the control plane with appropriate configuration and scale. This mechanism of extending the platform, too, follows the operator pattern, and you begin by introducing service-specific operators [30] and then instantiating these "add-on" services on top of the control plane. Appropriately privileged Kubernetes users just create and modify CRs to define what services to provision (or alter).

The platform technical stack

The Cloud Pak for Data stack, as shown in the following screenshot, is composed of four layers: the **Infrastructure-as-a-Service** (**IaaS**) layer at the bottom, and OpenShift Kubernetes providing the key compute, storage, and network infrastructure needed to support microservices on top.

IBM delivers the two key upper layers to this stack – **Foundational Services** that are leveraged as a common services layer by all IBM Cloud Paks. The Cloud Pak for Data control plane (called **Zen**), as the uppermost layer, manages the individual Cloud Pak for Data services and externalizes the API, as well as an integrated user experience:

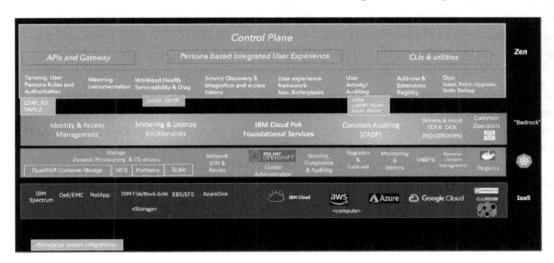

Figure 9.1 – Technical stack layers

In the following sections, we will dive deeper into the different layers of the stack and how Cloud Pak for Data, as a platform, powers data and AI solutions

Infrastructure requirements, storage, and networking

The IaaS layer could be on-premises or on the public cloud hyper-scalers. Compute could be virtualized, such as with VMware/vSphere or via Amazon AWS EC2 machines, Azure VMs, and so on, or physical bare-metal machines. These hosts form the Kubernetes master and worker nodes. Storage solutions could be native to the cloud (such as AWS EBS or IBM File Gold) or, in the case of on-premises deployments, leverage existing investments in storage solutions (such as IBM Spectrum Scale or Dell/EMC Isilon).

With OpenShift Container Platform v4, these hosts typically run **Red Hat Enterprise Linux CoreOS (RHCOS)**, a secure, operating system purpose-built for containerized workloads. It leverages the same Linux kernel, the same packages as the traditional RHEL installation, besides being designed for remote management. In addition, for improved security, it enables SELINUX out of the box and is deployed as an immutable operating system (except for a few configuration options).

OpenShift Container Platform can thus manage CoreOS Machine configurations, save states, and even recreate machines as needed. OCP's update process can also enable the continuous upgrade of the RHCOS hosts, making it much easier to maintain and operate clusters. RHCOS, with a smaller footprint, general immutability, and limited features helps present a reduced attack surface vector, improving the platform's overall security posture.

The Kubernetes community introduced a specification called the **Container Runtime Interface (CRI)** [8], which enables it to plug in different container engines. RHCOS includes the **CRI-O** [9] container engine, which enables integration with **Open Container Initiative (OCI)**-compliant images. CRI-O offers compatibility with Docker and OCI image formats and provides mechanisms to trust image sources, apart from managing the container process life cycles.

The following diagram provides a simplistic high-level outline of a Kubernetes cluster [*10*] to illustrate how it functions, just to set the context for how Cloud Pak for Data leverages it:

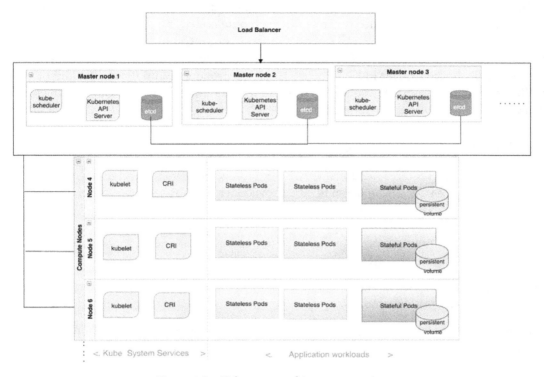

Figure 9.2 – Kubernetes architecture overview

There are essentially two types of host nodes – the **Master** and **Compute** (worker) nodes. In some implementations, a node may have both the master and compute roles, but it is highly recommended to maintain separate masters to ensure a better class of reliability. Each node includes some fundamental components [*11*] such as the Kubelet daemon and the functionality to interact with the container runtimes on each node. The master nodes include certain special components, such as the API server and the scheduler.

Kubernetes uses the etcd database for storing configuration information and this database is critical for overall operations. etcd is a key-value distributed store and, for high availability, can work with multiple replicas on different hosts, establishing a quorum. It is recommended that you start with three master nodes and expand when it becomes necessary to support a larger set of applications at scale.

The compute nodes are where application workloads are scheduled by Kubernetes to run. All Cloud Pak for Data Pods only run on the compute (worker) nodes. Kubernetes also includes the ability to schedule Pods in specific (dedicated) nodes as well, but in general, Pods can be randomly scheduled on any of the compute nodes. If a compute node, goes down, Pods could be rescheduled to run on a different node. Cloud Pak for Data services primarily leverage Kubernetes constructs such as Deployments [12] and StatefulSets [13] to schedule its microservices in Kubernetes. These "controllers" set up the right count of replicas for that microservice, identify what persistent volume storage to mount for the Pods, and declare requests for compute resources.

Understanding how storage is used

Cloud Pak for Data requires the use of a dynamic storage provisioner [20] in Kubernetes to create persistent volumes on demand. The provisioner also abstracts the use of different storage technologies and solutions in the background and makes Cloud Pak for Data portable. For example, an `nfs-client` storage class could point to an IBM Spectrum Scale or Dell/EMC-powered NFS server, while the `ibmc-file-gold-gid` storage class may point to storage in the IBM Cloud.

Hence, as part of the installation of Cloud Pak for Data, the user would provide storage classes as configuration parameters.

The Cloud Pak for Data control plane introduces a special persistent volume claim called `user-home-pvc`. This is a shared, **Read-Write-Many** (accessMode: RWX) volume that is mounted by several Pod replicas to access the same content. This volume stores some configuration information, static user experience content, and some user-introduced customizations.

In a typical Cloud Pak for Data installation, you can inspect your PVC as follows:

```
kubectl get pvc user-home-pvc
NAME             STATUS        VOLUME
CAPACITY     ACCESS MODES     STORAGECLASS     AGE
user-home-pvc    Bound         pvc-44e5a492-9921-41e1-bc42-
b96a9a4dd3dc     10Gi          RWX              nfs-client     33d
```

Looking at one of the deployments (such as `zen-core`), you will notice how the Deployment indicates where the volume should be mounted:

```
kubectl get deployment zen-core -o yaml
:
  - name: user-home-mount
      persistentVolumeClaim:
        claimName: user-home-pvc
:
volumeMounts:
    - mountPath: /user-home
      name: user-home-mount
```

In this example, you will notice that `user-home-pvc` gets mounted as `/user-home` in the `zen-core` Pods.

Cloud Pak for Data also uses StatefulSets, where each member replica is typically associated with its own dedicated (non-shared) persistent volume. Here's an example:

```
kubectl get statefulsets
```

This identifies a couple of StatefulSets present in the Cloud Pak for Data control plane:

```
NAME             READY   AGE
dsx-influxdb     1/1     33d
zen-metastoredb  3/3     33d
```

Each StatefulSet spawns one or more member Pods:

```
kubectl get pods -lcomponent=zen-metastoredb
```

The preceding snippet shows the *members* associated with that `metastoredb` StatefulSet. In this case, there are three of them and they are numbered ordinally:

```
NAME               READY   STATUS    RESTARTS   AGE
zen-metastoredb-0  1/1     Running   11         12d
zen-metastoredb-1  1/1     Running   11         12d
zen-metastoredb-2  1/1     Running   10         12d
```

Each StatefulSet member is typically associated with at least one persistent volume that serves to store data associated with that Pod. For example, each `zen-metastoredb` member is associated with a `datadir` volume where it stores its data. (The `zen-metastoredb` component is a SQL repository database that replicates data and responds to queries by load balancing three different members):

```
kubectl get pvc -lcomponent=zen-metastoredb
NAME                        STATUS    VOLUME
CAPACITY    ACCESS MODES    STORAGECLASS    AGE
datadir-zen-metastoredb-0   Bound     pvc-66bbec45-50e6-4941-
bbb4-63bbeb597403    10Gi     RWO          nfs-client
33d
datadir-zen-metastoredb-1   Bound     pvc-f0b31734-7efe-40f4-
a1c1-24b2608094da    10Gi     RWO          nfs-client
33d
datadir-zen-metastoredb-2   Bound     pvc-83fdb83e-8d2b-46ee-
b551-43acac9f0d6e    10Gi     RWO          nfs-client
33d
```

In *chapter 11, Storage* we will go into more detail on how storage is provided to application workloads in Kubernetes, as well as considering performance and reliability considerations.

Networking

Cloud Pak for Data leverages the concept of Kubernetes services [14] to expose these containerized microservices. A Kubernetes service (`svc`) exposes an internal hostname and one or more ports that represent an endpoint. A request to such a service will be routed to one or more Pod replicas in a load-balanced manner. Cloud Pak for Data services interact with each other by using the *cluster-internal* network enabled by Kubernetes and by using the name of the Kubernetes service. DNS enables Kubernetes services to be looked up by hostname and provides the fundamentals for service discovery. Take the following code snippet, for example:

```
kubectl describe svc zen-core-api-svc
```

This provides information relating to one microservice in Cloud Pak for Data's control plane. This `svc` (Kubernetes service) points to Pods identified by a specific set of `Selector` labels, for example:

```
Selector:         app.kubernetes.io/component=zen-core-
api,app.kubernetes.io/instance=0020-core,app.kubernetes.io/
```

```
managed-by=0020-zen-base,app.kubernetes.io/name=0020-zen-
base,app=0020-zen-base,component=zen-core-api,release=0020-core
```

`svc` also identifies which port is exposed and the protocol used, such as the following:

Port:	zencoreapi-tls 4444/TCP
TargetPort:	4444/TCP
Endpoints:	**10.254.16.52:4444,10.254.20.23:4444**

`Endpoints` indicate the Pod IPs and `4444` ports that the requests are routed to.

If you look at the list of `zen-core-api` Pods, you will notice that the IPs are assigned to those specific Pods:

```
kubectl get pods -lcomponent=zen-core-api -o wide
NAME                          READY    STATUS     RESTARTS    AGE
IP              NODE                            NOMINATED
NODE    READINESS GATES
zen-core-api-8cb44b776-2lv6q  1/1      Running    7           12d
10.254.20.23    worker1.zen-dev-01.cp.fyre.ibm.com    <none>
<none>
zen-core-api-8cb44b776-vxxlh  1/1      Running    10          12d
10.254.16.52    worker0.zen-dev-01.cp.fyre.ibm.com    <none>
<none>
```

Hence, in the case of a request to the `kube-svc` *hostname* `zen-core-api-svc`, port `4444` will be routed to one of these Pods automatically by Kubernetes. All other components within the same cluster would only use this `kube-svc` hostname to invoke its endpoints and are thus abstracted from the nature of the `zen-core-api` Pods, where exactly they run, or even how many replicas of these Pods may exist at the same time.

The Cloud Pak for Data control plane also runs a special service called `ibm-nginx-svc` that routes to `ibm-nginx` Pods. These Pods are delivered via a Deployment called `ibm-nginx`. This component, powered by nginx, is also known as the "front door" since it serves as the primary access point to Cloud Pak for Data services in general. Apart from any initial authentication checks, the `ibm-nginx` Pods also enable a reverse proxy to other microservices (identified by their `kube-svc` hostname) and serve the user experience.

OpenShift supports multiple ways [16] of accessing services from outside the cluster. Cloud Pak for Data exposes just *one* OpenShift route [17], called cpd, to invoke the user experience from a web browser and for invoking APIs from other clients:

```
kubectl get route cpd
NAME     HOST/PORT                                        PATH
SERVICES         PORT                    TERMINATION
WILDCARD
cpd     cpd-cpd-122.apps.zen-dev-01.cp.fyre.ibm.com
ibm-nginx-svc    ibm-nginx-https-port    passthrough/Redirect
None
```

This route represents the external access to Cloud Pak for Data and the "host" is a DNS resolvable hostname. In the preceding example, zen-dev-01.cp.fyre.ibm.com is the hostname of the OpenShift cluster itself. When a client from outside the cluster, such as a web browser, accesses this hostname, OpenShift routes it to the ibm-nginx-svc Kubernetes service, which is load-balanced across how many ibm-nginx Pods may currently be running.

OpenShift routes can also be secured using TLS [18] in general. Cloud Pak for Data also provides a way to associate custom certificates with its Kubernetes service [19].

Foundational services and the control plane

IBM has a broad portfolio of products that are now available as Cloud Paks on OpenShift Container Platform. With every such product, there is a common need to operate in different data centers, both on-premises and in the cloud, and the requirement to integrate with existing enterprise systems. To that end, a set of shared services has been developed called the **Cloud Pak foundational services (CPFS)**.

Cloud Pak foundational services

CPFS is also known as "Bedrock." It provides key services that run on top of OpenShift Container Platform to power all Cloud Paks, including Cloud Pak for Data, and serves as a layer to support integration between all Paks.

These capabilities enabled by the CPFS services include the following:

- **Certificate Management Service:** This service enables Cloud Paks to generate TLS certificates, manifest them as Kubernetes secrets, and be mounted in Pods that need them. This helps in securing inter-microservice communications and promotes easier automatic rotation of certificates.

- **Identity and Access Management (IAM) service**: IAM enables authentication for the Cloud Pak. It provides mechanisms for admins to configure one or more identity providers (such as LDAP/AD) to help integrate with existing systems in the enterprise. IAM exposes the **OpenID Connect (OIDC)** standard that also makes it possible for users to single sign-on between multiple Pak installations.

- **License and metering service**: This service captures utilization metrics for individual services in the cluster. Each Cloud Pak Pod is instrumented with details that identify it uniquely. The license service collects the **Virtual Processor Core (VPC)** resources associated with each Pod and aggregates them at a product level. It also provides reports for license audit and compliance purposes.

- **Operand Deployment Lifecycle Manager** (**ODLM**): The operator pattern is used by all Cloud Paks to deliver and manage services on OpenShift Kubernetes. Kubernetes resources, such as Deployments and StatefulSets that are orchestrated and managed by an operator, are collectively referred to as the *Operand* controlled by that operator. ODLM [*21*] is an operator, developed as an open source project, that is used to manage the life cycle of such operands, in a similar way to how OLM manages operators. With higher-level constructs such as *OperandRequests*, ODLM can provide abstractions and simpler patterns for Paks to manage operands and even define inter-dependencies between them.

- **Namespace Scoping and Projection of privileges**: Operators are granted significant **Role-Based Access Control (RBAC)** privileges to perform their functions. They connect to the Kubernetes API server and invoke functions that allow it to deploy and manage Kubernetes resources in the cluster. It is very common for OpenShift clusters to be shared among multiple tenants and multiple vendors. Hence, from a security perspective, it becomes imperative to be able to control the breadth of access that both operators and operands are granted.

Frequently, operands are assigned to specific tenant namespaces and are expected to only operate within that namespace. Operators, however, are installed in a central namespace and need to have authority in these individual tenant namespaces. At the same time, it is desirable not to grant cluster-wide authority to operators and to limit their influence only to those namespaces.

IBM introduced the namespace scope operator [*22*], in open source, to help address this need. This operator can *project* the authority of operators (and operands if needed) to other namespaces selectively. It also provides the ability to automatically change the configurations of these operators to have them watch additional namespaces and only those namespaces, thereby improving the security posture of the Paks in shared clusters.

- **Zen – the platform experience**: Zen is a framework developed specifically to enable the extensibility and adoption of a plug-n-play paradigm for both the end user experience and for backend service APIs. It is a set of foundational capabilities and UI components that are leveraged by higher-level services in Cloud Paks. It ensures a focus on a single pane of glass for *persona-driven* and customizable (even re-brandable) user experiences. It fosters collaborations between personas and across product boundaries and, with the plug-n-play model, capabilities are dynamically enabled as and when services are provisioned.

- **The cloudctl utility**: This **command-line interface** (**CLI**) provides functions to deploy and manage Cloud Paks on OpenShift Container Platform. cloudctl helps with mirroring Cloud Pak images into an Enterprise Private Container Registry, even into air-gapped environments. Services in Cloud Paks are packaged and made available in the **Container Application Software for Enterprises** (**CASE**) format, a specification defined in open source [*23*]. The utility enables important automation for installing and upgrading services in a Pak, including creating OLM artifacts such as catalog sources and operator subscriptions.

In this section, we introduced the functionality that the Cloud Pak foundational services deliver, forming the "Bedrock" that is leveraged by all IBM Cloud Paks. In the next section, we will explore one such consumer of Bedrock – the *Control Plane* layer that powers Cloud Pak for Data services.

Cloud Pak for Data control plane

The control plane is an instantiation of the Zen framework in Cloud Pak for Data instance namespaces. It is a set of deployments that form the operand managed by the Zen operator. This Zen control plane is introduced and customized by the Cloud Pak for Data platform operator [*5*].

The following screenshot provides an architectural overview of the different Kubernetes resources. It depicts the various microservices, repositories, storage volumes, and other components that make up the control plane. It also shows service-to-service interactions, including with Cloud Pak foundational services and operators, as well as how web browsers and command-line utilities interact with that installation of Cloud Pak for Data:

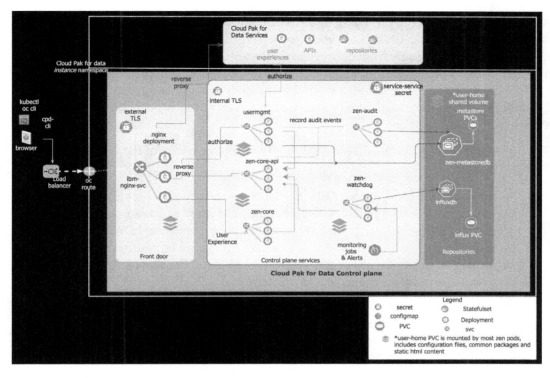

Figure 9.3 – Control Plane components and interactions

The control plane is powered by an extension registry, which is an API (delivered by the zen-core-api microservice) and a repository (**zen-metastoredb**) that enables a dynamic plug-n-play of services and other components. The control plane's **zen-core-api** microservice hosts key services and APIs, to enable service discovery and enforce authorizations, and helps present a dynamic, integrated experience to the end user.

As described earlier, the ibm-nginx microservice serves as the "front door" access point for both the user experience and API invocations. In the preceding screenshot, you will notice that the operators located in the Cloud Pak foundational services namespace manage the "operands" in the Cloud Pak for Data instance namespace.

As coarsely outlined in the preceding figure, all the Cloud Pak for Data services are deployed on top of the control plane, and they leverage the control plane APIs as well as get accessed via the same front door. These higher-order services themselves are managed by their own operators, include their own microservices, repositories, and user experiences over and above what the control plane itself hosts. These service experiences dynamically extend the Cloud Pak for Data experience in a plug-n-play fashion to present a single pane of glass for end users.

The control plane provides the following capabilities for consumers of the Cloud Pak for Data platform:

- **Authorization**: Cloud Pak for Data provides an access control capability (the *usermgmt* microservice) that enables an authorized user (an administrator) to grant access to other users. User groups, which can also be mapped to LDAP groups, can be defined to support collaboration. Users and groups can be granted role-based authorization to work with specific services or projects. The chapter on security describes user access management in greater detail.

- **A persona-based experience**: Cloud Pak for Data provides a browser-based integrated end user experience that is personalized appropriately for the user's role. Navigation and their home page content are automatically scoped to the access rights that have been granted. Users can choose to customize their experience as well, say with the home page content. This is delivered primarily by the "zen-core" micro-service.

- **A service catalog**: The service catalog encourages a self-service option for authorized users to *provision* instances of services in the Cloud Pak for Data platform without needing to have any sophisticated knowledge about the underlying Kubernetes infrastructure. For example, a user could choose to provision a Db2 database and grant access to others to work with that database instance. Cloud Pak for Data also includes mechanisms for users to generate API keys for service instances to programmatically interact from client applications. The catalog is powered by the `zen-core-api` microservice and the experience is hosted by the `zen-core` microservice. `zen-core-api` also provides the mechanism to manage access to each provisioned service instance, while supporting the ability to upgrade these instances as needed.

- **Auditing**: The control plane includes a `zen-audit` service that is used to control and abstract how auditing-related events are to be delivered to a collector in the enterprise. Microservices, whether in the control plane itself or those from Cloud Pak for Data services, emit audit events for privileged actions to this central audit service. The audit service also can also grab audit records that have been placed in special log files by individual microservices. The chapter on security describes how auditing for compliance can be supported across the whole stack, beyond just the Cloud Pak for Data platform layers.

- **Monitoring**: The Cloud Pak for Data control plane includes a service called the "zen-watchdog" that provides the basic capability to monitor the control plane itself and all the Cloud Pak for Data services on top of it. The next section describes this in more detail.

- **The cpd-cli utility**: This utility is a command-line program [24] that provides key capabilities (introduced as plugins) to manage the control plane and leverage capabilities from a client system. For example, an administrator could leverage this program to bulk authorize users for access or to gather diagnostics, or simply to work with service instances that this user has access to. The backup-restore plugin enables `cpd-cli` to support additional use cases.

> **Note**
> With Cloud Pak for Data v3 and v3.5, this utility was used to support Cloud Pak for Data service installs, patches, and upgrades as well. However, with the use of operators, these functions are now driven by the service operators and by specialist fields in the appropriate custom resources.

Management and monitoring

Cloud Pak for Data provides APIs and a user experience for administrators to manage the installation of Cloud Pak for Data.

> **Note**
> This does not require that these users are granted Kubernetes access either, only appropriate admin roles/permissions in Cloud Pak for Data. Such users are also referred to as Cloud Pak for Data *platform* administrators and would only have authority within that single instance of Cloud Pak for Data.

User access management, monitoring, configuration, and customizations, as well as the provisioning of services or storage volumes, are some of the typical administration functions in Cloud Pak for Data. In this section, we will look at some of these. Other chapters (such as *Security for User Access Management*) will dive deeper into some of these aspects.

Monitoring

The Cloud Pak for Data control plane provides APIs and user experiences for authorized end users to be able to monitor and control workloads in that installation.

The zen-watchdog service leverages labels and annotations [25] in different Pods to identify the service they are associated with and presents a user experience for administrators. The zen-watchdog service also periodically spins off Kubernetes Jobs to capture metrics associated with these individual Pods, aggregates them, and stores them in the influxdb time series databases.

The following screenshot shows the **Monitoring** page, which provides an at-a-glance view of the state of that installation:

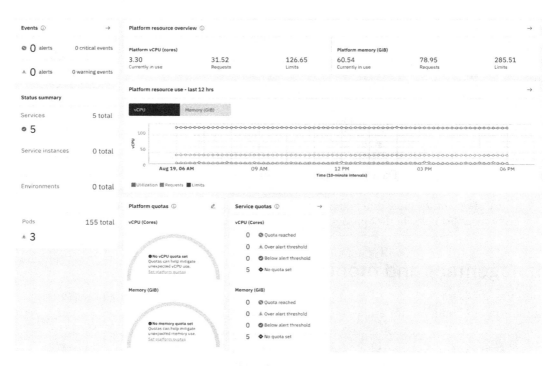

Figure 9.4 – Monitoring overview

This page also reveals any events that need attention and summarizes recent resource consumption. Users can drill down to see what services have been deployed and their overall resource consumption, as shown in the following screenshot:

Figure 9.5 – Status and resource consumption by service

The preceding screenshot shows a user experience that helps users see a breakdown by service. It presents a view of the overall status of that service and resource utilization metrics. Users can then drill down further to get advanced details about one specific service.

The following screenshot, as an example, shows the specific Pods associated with the Watson Knowledge Catalog Service. A similar experience exists to show all Pods from all services as well, with the ability to sort and filter to quickly identify Pods that need attention:

Pod name	Service	Current vCPU use	vCPU requests	Current memory use (GiB)	Memory requests (GiB)
Total		3.79	31.52	60.44	78.95
shop4info-mappers-service-0	Watson Knowledg...	0.01	0.20	0.29	0.50
shop4info-rest-0	Watson Knowledg...	0.00	0.20	0.48	1.00
shop4info-scheduler-0	Watson Knowledg...	0.00	0.10	0.33	1.00
shop4info-type-registry-servic...	Watson Knowledg...	0.00	0.10	0.32	0.50
solr-0	Watson Knowledg...	0.01	0.15	0.84	1.00
spaces-7556bf4568-5jz7w	Common Core Ser...	0.01	0.50	0.16	0.50
spawner-api-54bfc796d5-hx6j2	Common Core Ser...	0.00	0.01	0.03	0.06
usermgmt-5b79c55dd9-5j49b	IBM Cloud Pak for...	0.01	0.20	0.09	0.25
usermgmt-5b79c55dd9-kr7j6	IBM Cloud Pak for...	0.00	0.20	0.09	0.25
wdp-activities-85	Watson Knowledg...	0.06	0.10	0.51	0.78
wdp-connect-con	Common Core Ser...	0.02	0.15	1.40	0.63
wdp-connect-con	Common Core Ser...	0.02	0.20	1.40	0.63
wdp-connect-flig	Common Core Ser...	0.01	0.20	1.50	0.63
wdp-couchdb-0	Common Core Ser...	0.11	1.00	1.80	0.50
wdp-couchdb-1	Common Core Ser...	0.11	1.00	1.90	0.50
wdp-couchdb-2	Common Core Ser...	0.11	1.00	1.90	0.50

Figure 9.6 – Pod details and actions

From this user experience, as shown in the preceding screenshot, the administrator can then look at each Pod, get more details, or access logs for troubleshooting. They could even choose to restart a Pod if they see it malfunctioning.

Note that this interface is meant to complement what is available in the OpenShift console for monitoring and management, and only presents a view that is very much in Cloud Pak for Data's usage context and only scoped to that Cloud Pak for Data installation.

Alerting

Kubernetes provides the concept of liveness probes and other primitives for it to recognize failing Pods and trigger automatic restarts. However, with complicated services, it may not be that simple to recognize failures that span multiple deployments or StatefulSets, nor would it be able to recognize any application-level semantics. To complement OpenShift's capabilities, Cloud Pak for Data introduces additional monitors, mechanisms to persistent events, and surface issues in the user experience.

The control plane includes a framework [27], powered by zen-watchdog, to monitor for events, recognize serious situations, and raise those problems as alerts. These alert messages are frequently sent as SNMP traps to configured Enterprise SNMP receivers.

The following diagram provides a high-level view of how the zen-watchdog service enables monitoring, the aggregation of events, and the raising of alerts:

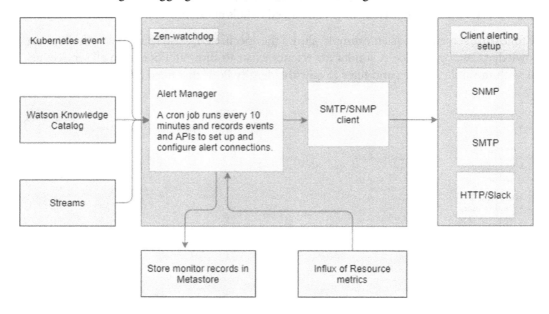

Figure 9.7 – Monitoring events and alert forwarding

Different targets can be configured to send alerts to, including via the SNMP protocol, or simply as SMTP emails as well as via HTTP.

Cloud Pak for Data's control plane itself includes a few generic out-of-the-box monitors, such as the following:

- **check-pvc-status**: Records a critical event if a PVC is unbound
- **check-replica-status**: Records a critical event if a StatefulSet or Deployment has unavailable replicas
- **check-resource-status**: Records a warning event if a service has reached the threshold and a critical event if it has exceeded the quota

Developers can introduce extensions to support custom monitoring needs as well. The documentation [28], using sample code, describes the procedure of introducing a custom container image to provide specialized monitoring functions and identify critical problems that merit an alert.

Provisioning and managing service instances

Many services in Cloud Pak for Data support the concept of "service instances," provisioned from the service catalog. These are individual copies of services that can be scaled out and configured independently, or often used to provide access to a selected group of users.

A service instance has the following characteristics:

- It represents Kubernetes workloads that are metered individually – usually provisioned in the same namespace as the control plane – but some services can also be provisioned in a "data plane" sidecar namespace that is *tethered* (from a management perspective) to the control plane namespace.
- It is provisioned by authorized end users with *Create service instances* permission via a provisioning experience and API:

  End users provision instances without needing to have Kubernetes access or an in-depth understanding of Kubernetes primitives.
- It supports RBAC, in other words, users can be granted granular roles to access that resource.
- It enables secure programmatic client access, with instance-scoped API keys and service tokens.

For example, an authorized user may want to spin up an instance of a database for us by their team privately. Other users would not even be made aware that these instances even exist.

The following screenshot shows a sample provisioning experience to deploy a Db2 warehouse database instance.

The user is guided on setting up the instance, including assigning compute and storage appropriately:

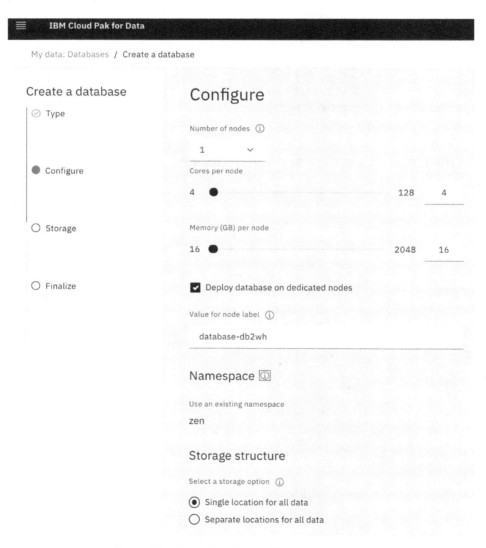

Figure 9.8 – Example of a provisioning experience

After choosing the appropriate attributes for provisioning that instance, for example, as shown in the preceding screenshot, the user then triggers the creation of the instance. The creator of such an instance is then designated as the first administrator of that instance.

With the creation of such an instance (or generally whenever any specific operations are performed on that instance), audit records are automatically generated by the control plane.

Cloud Pak for Data platform administrators, apart from the creator of the instance, can also manage and monitor these instances. However, even platform administrators may be denied access to consume the instance in the first place. For example, while the Cloud Pak for Data platform administrator can perform maintenance tasks, or even de-provision that database instance, they may not have access to the data inside. This is critical from a regulatory compliance perspective to enable such separation of duties.

The following screenshot shows an example of how an authorized user (the admin of that instance or the Cloud Pak for Data platform administrator) can manage access rights. They can assign roles (including that of an admin) to other users and groups:

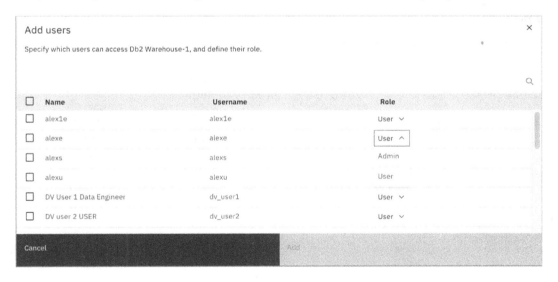

Figure 9.9 – Instance access management

It is also common for the platform administrator to retain the *Create Service instance* permission. These would be the only authorized users to provision such instances. However, once the instances have been provisioned, they can then grant the admin role on that instance to other users or groups of users, essentially delegating the day-to-day management of that instance. Newly added instance admins could then remove access from platform administrators (creators) to ensure compliance. The platform administrators can still rely on the monitoring interfaces to watch over such instances or diagnose and correct problems.

The `cpd-cli` service-instance command can also be used to manage such instances from the command line.

Storage volumes

The chapter on storage describes some of the fundamentals behind how persistence is enabled in Kubernetes and how Cloud Pak for Data leverages those. In this section, we will focus on a capability called **storage volumes**, a mechanism by which administrators enable end users to access shareable storage for the implementation of their use cases on the platform.

Authorized users, typically Cloud Pak for Data administrators, can provision additional persistent volumes or make available remote volumes for consumption by users in the platform [*32*]. This is usually for sharing data or any source code packages between different project teams and across different Kubernetes containers.

For example, a storage volume could be provisioned afresh in the cluster and end users, or groups of users could be granted access to that storage volume. These users can then access them inside projects, for example, inside multiple Jupyter or RStudio environment Pods, or for executing a Spark Job or generally via APIs [*33*].

The Cloud Pak for Data administrator can provision and manage such volumes via the user experience, as shown in the following screenshot:

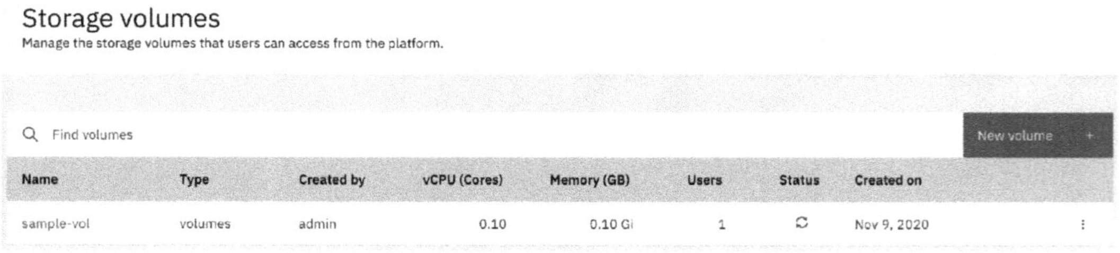

Storage volumes
Manage the storage volumes that users can access from the platform.

Name	Type	Created by	vCPU (Cores)	Memory (GB)	Users	Status	Created on	
sample-vol	volumes	admin	0.10	0.10 Gi	1	↻	Nov 9, 2020	⋮

Figure 9.10 – Managing storage volumes

The Storage volumes page enables users to view all the volumes they have access to as well as browse through the files in those volumes. The experience also supports rudimentary file uploads and downloads as well from these volumes.

Volumes are resources, too, and since they are likely to contain important data, the Cloud Pak for Data control plane provides mechanisms for the "owner" of the volume to grant access to other users and groups, as shown in the following screenshot:

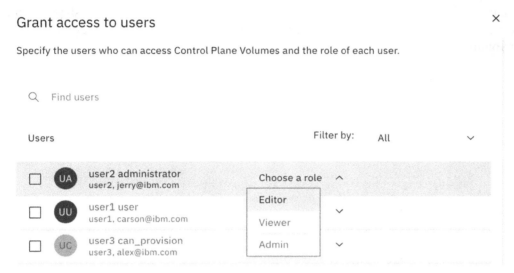

Figure 9.11 – Controlling access to a storage volume

Users may even be granted "admin" rights to an individual volume, allowing them full authority to the volume, including managing access and de-provisioning the volume. Others may be allowed write access, and others still just read access.

Cloud Pak for Data supports different ways of providing such shareable storage volumes, as shown in the following screenshot:

Storage volume details

Volume type ⓘ

Select type   ∧

External NFS

External SMB

Existing PVC

New PVC

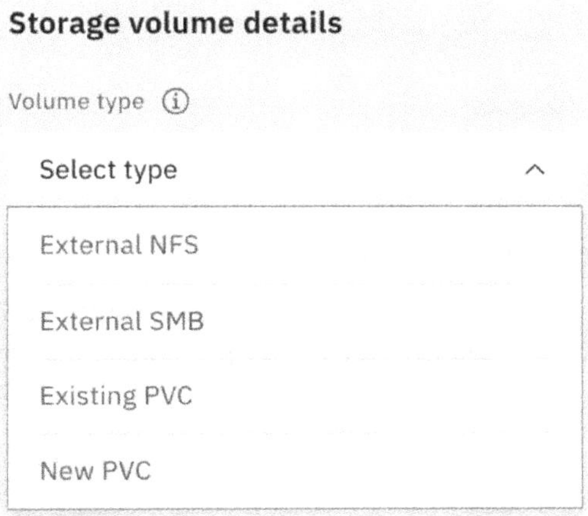

Figure 9.12 – Types of storage volumes

An administrator could decide to create a fresh volume using an existing Kubernetes storage class (RWX access mode) or point to an existing persistent volume or mount remote volumes using either the NFS or SMB protocols.

As an example, the following screenshot shows an administrator creating a fresh persistent volume using the cephfs storage class and identifying the expected mount point for that volume:

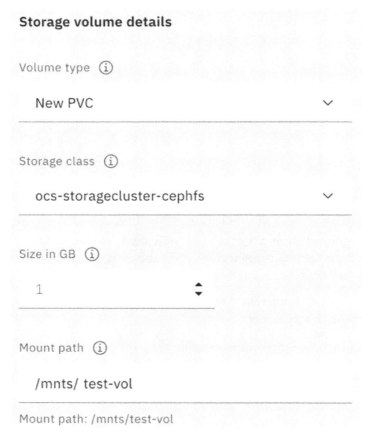

Figure 9.13 – Creation of a new storage volume

Note that the storage class must support the **Read-Write-Many** (**RWX**) access mode since these volumes are expected to be mounted by multiple Pods concurrently.

The following screenshot illustrates how a remote volume can be made available to users in a similar fashion:

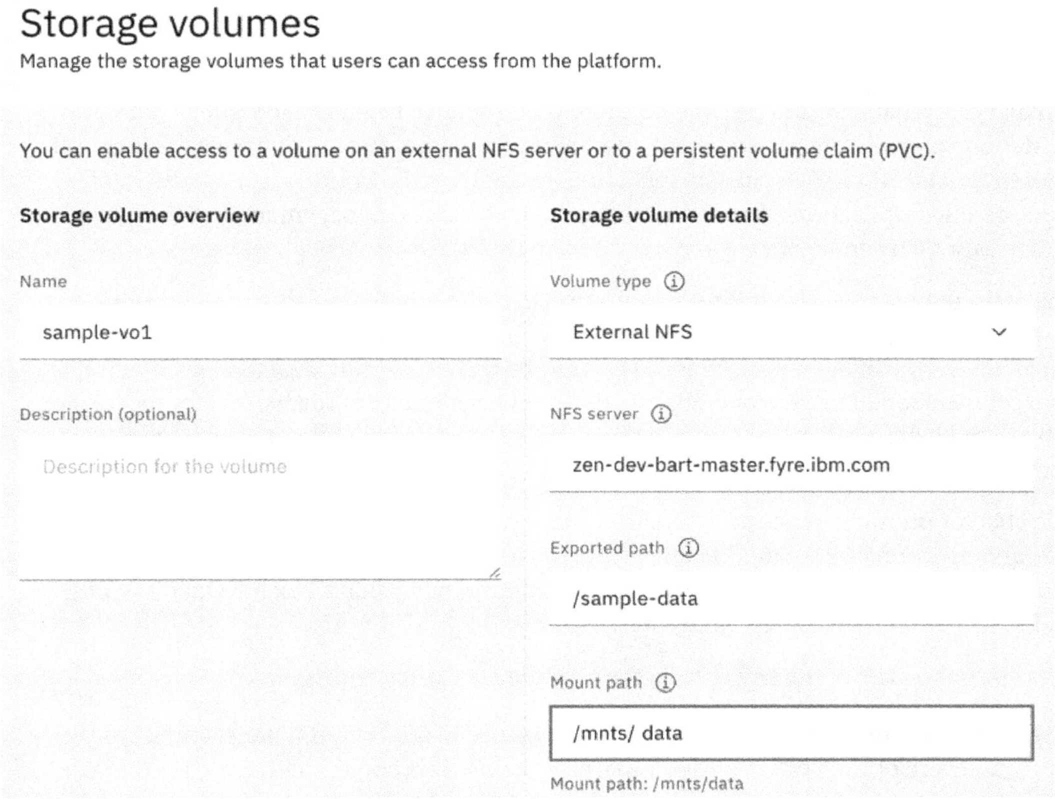

Storage volumes

Manage the storage volumes that users can access from the platform.

You can enable access to a volume on an external NFS server or to a persistent volume claim (PVC).

Storage volume overview

Name

sample-vo1

Description (optional)

Description for the volume

Storage volume details

Volume type (i)

External NFS

NFS server (i)

zen-dev-bart-master.fyre.ibm.com

Exported path (i)

/sample-data

Mount path (i)

/mnts/ data

Mount path: /mnts/data

Figure 9.14 – Accessing a remote volume

This enables enterprises to use existing data volumes inside Cloud Pak for Data without the need to physically transfer them into the cluster. Similarly, the SMB protocol is also supported to mount remote SMB (CIFS) volumes, similar to how Microsoft Windows machines mount such shared volumes.

Multi-tenancy, resource management, and security

OpenShift clusters are considered shared environments in many mature enterprises. They expect to expand such clusters when needed to accommodate more workloads and re-balance available resources. It is far more cost-effective to share the same OpenShift cluster and consolidate management and operations rather than assign a separate cluster to each tenant. Since such clusters include different types of workloads from different vendors, and applications that require sophisticated access management and security practices, a focus on tenancy is important from the start.

The concept of tenancy itself is very subjective. In some situations, dev-test and production installations may be considered different tenants in the same cluster. In other cases, different departments, or different project use cases, may be treated as individual tenants. When this concept is extended to ISVs operating a cluster on behalf of their clients, each of those clients is likely to be different companies and there could be stricter requirements on tenancy compared with a case where all users are from the same company.

The chapter on multi-tenancy covers the tenancy approach and best practice recommendations with Cloud Pak for Data in greater detail. In this section, we will look at a high-level overview of tenancy requirements and how Cloud Pak for Data is deployed and managed in such shared Kubernetes clusters.

Let's first start with some key criteria that influence tenancy approaches:

- **Isolation of tenants**: There is a need to ensure that different tenants are fully unaware of each other on such shared systems. This could include the need to use completely different authentication mechanisms. Different LDAP groups or even different LDAP servers may be desired for different tenants.

  Protecting resources via access authorizations and policies is absolutely necessary in general, but with multiple tenant users present, these assume an even more prominent role.

  Isolation also extends to network isolations, even to the point of having completely different user experience and API hostname URLs and DNS domain names for each tenant, even if they physically share the same Kubernetes cluster.

  Storage systems, consumed in the form of Kubernetes persistent volumes, also need to ensure that tenants do not have access to each other's volumes.

- **Security**: Note, however, that for cost reasons, some enterprises may end up permitting a lot more sharing between tenants. So, while isolation requirements provide some amount of separation between tenants, there are also specific security and regulatory compliance requirements that play a fundamental role in the Deployment topology. Authentication and authorization access management help to a great extent, but may not be sufficient.

  For example, since OpenShift Kubernetes itself is shared, it would be unwise to grant access to any tenant to that operating environment. This also means that there needs to be an operations team or IT department solely tasked with managing the OpenShift cluster, including the security of the host nodes themselves, as well as the onboarding of tenants.

  As with traditional, non-containerized applications, there could also be a need for network security. For example, two tenants could need to be firewalled off from each other or strict policies set to ensure that no breach occurs.

  Security requirements could extend to even needing different encryption keys for storage volumes for different tenants.

  From a regulatory compliance perspective, different tenants may have different requirements. There may not be any need to have auditing for dev-test or ephemeral tenants, but more thoroughness may be required for production tenants. Some tenants, because of the nature of their work with sensitive data, may require a lot more security hardening and audit practices than others. Hence, there is a need for the tenancy approach to support such diverse configuration requirements as well.

- **Resource management**: With a shared cluster offering a pool of compute, it is also imperative that the cluster is shared equitably. IT operation teams may need to support specific **Service Level Agreements (SLAs)** for individual tenants.

  Hence, the tenancy mechanism must support controls where resource utilization is monitored, noisy neighbors are prevented or at least throttled, and importantly, compute usage limits can be enforced. In some organizations, chargebacks to tenants could also rely on the ability to measure compute utilization and/or set quotas.

- **Supporting self-service management**: It becomes operationally expensive for enterprises to have dedicated support for each tenant user. It is desirable to have each tenant self-serve themselves, even for provisioning services or scaling them out, or performing backups or upgrades on their own.

It is also usually the case for a tenant admin to decide on matters of access management (and compute resource quotas) on their own, without needing IT support.

It is preferable for monitoring and diagnostics or other maintenance tasks to be delegated to the tenant admins as well.

Approaches to tenancy should thus be able to support granting some level of management privileges to designated "tenant administrators" within their scope.

In this section, we looked at some common requirements that are placed on multi-tenancy approaches. The next section will introduce how the Cloud Pak for Data technology stack helps address these aspects of tenancy.

Isolation using namespaces

While it may be simpler to use different Kubernetes clusters or even dedicated host nodes for each tenant, it will not be cost-effective. Luckily, Kubernetes includes the concept of a namespace [34] that represents a *virtual cluster* to help sandbox different tenant workloads. OpenShift Container Platform extends namespaces with the concept of a project [35].

The most common approach is to have different installations of Cloud Pak for Data in different tenant *instance* namespaces. Each tenant is then assigned one such installation and works in relative privacy and isolation from one another. Cloud Pak for Data installations in each tenant namespace use their own OpenShift router, which allows for the customization per tenant of this access point, including external hostnames and URLs, as well as the ability to introduce different TLS certificates for each tenant.

Since persistent volume claims and network policies can be scoped to namespaces as well, isolation from storage and network access can also be guaranteed. Should a tenant user need access to Kubernetes (for example, to install or upgrade their services), that could be scoped to their namespace, too, with Kubernetes RBAC. OpenShift projects could also be mapped to dedicated nodes and this can provide for more physical separation, at an increased cost, to isolate Pods from various tenants from one another.

The chapter on multi-tenancy describes the various options of using namespaces to achieve isolation as well as options where two tenants may want to even share the same namespace.

Resource management and quotas

Cloud Pak for Data supports a wide variety of use cases and different types of users concurrently using an installation. Once a service is deployed, there would be some fixed set of Pods that provide APIs, repositories, and user experiences, as well as some supporting management aspects specific to that service. These are considered *static* workloads that are always running. When end users interact with these services, there typically would be *dynamic* workloads. Some of these may be long-running, such as a user-provisioned database instance, and others may simply be short-term workloads such as data science interactive environments, Spark jobs, and machine learning training jobs. If there are more users concurrently launching analytics environments or, say, initiating expensive model training simultaneously, they may end up competing for resources or starving other essential jobs.

All Pods in Cloud Pak for Data are labeled and annotated to identify the *product* it belongs to [25]. Pods also specify resource requests and limits needed [29] for operating in the Kubernetes environment. The request CPU and memory settings help Kubernetes pick a suitable worker node to schedule the Pod in, reserve the requested compute for that Pod, and use the limit settings to set an upper limit. Pods would thus be able to "burst" to the upper limit for a short duration and be able to support a sudden spike in workloads. This avoids the need to plan for the worst case and reserve more compute than typically needed, while allowing emergency, temporary growth, but only if additional resources are available in that worker node at that instant.

Resource management is a critical aspect to consider when working with such dynamic workloads and concurrent user activity. While the architecture elegantly supports elasticity and more compute resources can be added when needed, it is more practical and cost-effective to fully consume existing resources to the maximum extent before scaling out. An administrator would thus need to ensure that there is a fair and balanced use of resources and can transfer resources to workloads that urgently need it.

Kubernetes supports the concept of a **ResourceQuota** specification that can be used by cluster administrators to control compute allocations across all namespaces. In addition to this, the control plane includes the ability for a Cloud Pak for Data *tenant* administrator, without needing Kubernetes privileges, to define compute resource thresholds per service (product) in their installation. While this is still subject to the overall namespace ResourceQuota set by the Kubernetes cluster admin, it allows some flexibility to the Cloud Pak for Data administrator to self-manage the resources that have been granted.

The following screenshot illustrates how such a tenant administrator can set boundaries on both CPU and memory for the Watson Knowledge Catalog Service within that Cloud Pak for Data instance namespace:

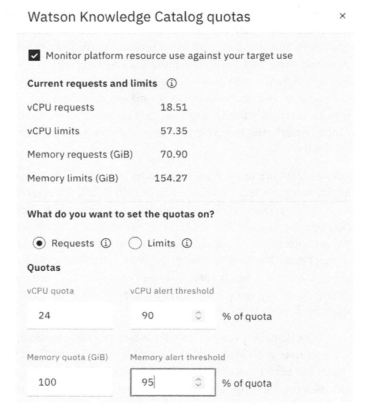

Figure 9.15 – Resource quota definition

Any violations on the thresholds are highlighted in the monitoring user experience as well as raised as alerts, if configured.

Beyond just reporting on threshold breaches, Cloud Pak for Data includes an optional scheduling service [26] that can be used to *enforce* assigned resource quotas. The scheduling service is configured based on the choices made by the administrator per service.

The scheduling service extends the default Kubernetes scheduler, and monitors all start up requests for Pods and Jobs. It uses the same labels that zen-watchdog also uses to identify the Pods and the service/products that the Pod belongs to. It then aggregates resource requests and limits across all Pods that are currently running for that service and decides whether to permit the startup. A Pod that would violate the defined resource quota threshold is prevented from even being started.

Enabling tenant self-management

The approach of using different namespaces for each tenant also elegantly solves the problem of self-management with Cloud Pak for Data.

Each Cloud Pak for Data installation, in its own namespace, can be independently configured and assigned to be administered by a tenant user. That tenant administrator has access to all the monitoring and alerting capabilities, in addition to resource management as described in the previous sections.

Each such tenant instance can have different Cloud Pak for Data services and versions deployed completely independently from other tenants. This also enables each tenant administrator to make a decision regarding upgrading and patching, too, on their own terms.

Day 2 operations

Administrators of Cloud Pak for Data installations, including tenant administrators, can leverage utilities and best practices when it comes to the management of the Cloud Pak for Data platforms and services.

This section describes some of the operational considerations beyond just the initial installation, daily management activities, and monitoring.

Upgrades

With the adoption of the operator pattern, Cloud Pak for Data makes it possible for even individual tenant user to decide on their own an upgrade strategy that works for them. In certain cases, Cloud Pak for Data instances may be upgraded more frequently, even to the latest major release versions, such as with dev-test installations, whereas in other instances, such as production installations, there is frequently a need to stick to stable releases and only incremental changes (patches) may be tolerated.

In general, there are two distinct sets of components that you would need to consider from an upgrade perspective – the *operators* in the central namespace, and the individual *operand* service deployments in (tenant) Cloud Pak for Data instance namespaces. The typical procedure involves upgrading all the operators to the latest released version (or just the operators of immediate interest) and then deciding which operands to upgrade. The operator pattern enables both a fully automated upgrade and selected manual upgrades or even a mix as appropriate for different tenants.

With the *installPlanApproval: Automatic* specification in the operator subscriptions [*36*], operator deployments are typically set to be automatically upgraded when its catalog source is refreshed, A choice of a manual approval implies that a cluster admin manually decides when the upgrade is acceptable. Note that new versions of Cloud Pak for Data operators are expected to still tolerate currently existing installed versions of the operand deployments and support new versions at the same time. Hence, in the same cluster, a fresh installation of the new version of the service could be deployed, while another namespace continues to retain the old version, and both get managed by the same operator.

Operands can be set to be auto-upgraded as well. This would mean that when the operator is upgraded, as part of its reconcile loop, it automatically upgrades existing operands as well to the latest version supported by that operator.

The easiest approach is to configure the fully automated mechanisms and adopt the continuous delivery paradigm. With this approach, all the cluster administrator needs to do is to simply ensure that the catalog source is updated (manually if air-gapped). Operators introduced by that updated catalog service would then get automatically upgraded to the latest version and those operators would, in turn, automatically upgrade the operands they manage.

For situations where a specific version of an operand is required and upgrades need to be handled manually, a process called "pinning" can be adopted. Pinning operands to a version works even when the operator concerned has been upgraded. This is done by explicitly providing a version in the custom resource spec for that service in an instance namespace. Only when the version is manually altered would the operator trigger the upgrade of the operand deployments to the new version requested. If no version is specified, then the operator assumes that the operand deployments always need to be at the latest release level and automatically ensures that to be the case.

Scale-out

Out of the box, operand deployments of services are configured to use specific compute resources and replica settings. Most services support a T-shirt sizing configuration that provides for the scaling out and scaling up of the various Pod microservices. In many cases, scale-out is also done to improve high availability with multiple replicas for deployments and improved performance with increased load balancing.

The Cloud Pak for Data documentation [37] describes the service-specific configuration options for scaling. For the scale-out/up operation, the tenant administrator updates a *scaleConfig* parameter in the appropriate custom resource spec. For example, a medium *scaleConfig* would cause that service operator to alter (and maintain) the configuration of the operand deployments to the size represented by that *scaleConfig*. This could mean specific resource requests and limits per Pod [29], as well as different replica counts in Kubernetes Deployments and StatefulSets.

It is not necessary to scale out all Cloud Pak for Data services at the same time or to the same configuration within the same Cloud Pak for Data installation. You would only need to scale out based on expected workloads and the concurrency of specific use cases. You could choose to scale back down some services and release resources to scale out other services for a different set of use cases later. Different tenant installations of Cloud Pak for Data could also have a different set of services at different scale configurations.

Backup and restore

Cloud Pak for Data relies on the storage solution in that Kubernetes cluster for reliable persistence of data. Storage solutions provide for redundancy as well with block-level storage replication. *Chapter 11, Storage* describes the technological aspects and reliability in general.

Cloud Pak for Data includes a backup-and-restore utility [38] as part of `cpd-cli` that helps Kubernetes namespace administrators take point-in-time backups of data and metadata from that Cloud Pak for Data instance. This allows restoration in the same installation to restore to a previously known state as well as restoration in a different cluster as part of a disaster recovery procedure.

The common practice is to trigger a backup of data from all persistent volumes in a particular OpenShift Project/Kubernetes namespace and to persist that backup in a remote volume or an S3-compatible object store. In some cases, the selected storage solution may be limited and may not be able to guarantee consistency of the backup images since it could span many persistent volumes. In such a case, cpd-cli also provides a way to "quiesce" the services to prevent writes to the volumes, to essentially set that installation to a maintenance mode, and an "unquiesce" once the backup procedure is complete. The backup is simply a copy of all the persistent volume filesystems [39]. One disadvantage here is that the *quiesce-backup-unquiesce* flow may take a while, leading to that installation staying in a longer than acceptable maintenance mode. In other cases, where the storage solution has the necessary sophistication [40], Cloud Pak for Data services may put I/O in a write suspend mode for a short duration while all buffers are flushed to disk. Writes are held back while storage *snapshots* are taken, resulting in negligible disruption compared to a full quiesce.

Summary

Cloud Pak for Data is a platform for data and AI, built on top of a modern Kubernetes-based architecture. It leverages OpenShift Container Platform and Kubernetes to deliver a reliable, elastic, and cost-effective solution for data and AI use cases.

In this chapter, you were introduced to how Cloud Pak for Data is built, the different technology layers that form the stack that powers it, and the roles that the Cloud Pak foundational services and the control plane play. We also discussed the relevance of the operator pattern from the perspective of installation of the services and upgrades. Tenancy requirements and management of Cloud Pak for Data by individual tenant administrators, as well as approaches to resource management, were introduced. Day 2 operations were outlined as well.

In subsequent chapters, we will dive deeper into some of the key aspects introduced in this chapter, including storage, security, and multi-tenancy.

References

1. **What is Kubernetes?**: https://kubernetes.io/docs/concepts/overview/what-is-kubernetes/
2. **Red Hat OpenShift Kubernetes for enterprises**: https://www.openshift.com/learn/topics/kubernetes/
3. **Operator pattern**: https://kubernetes.io/docs/concepts/extend-kubernetes/operator/
4. **Operators**: https://operatorframework.io/what/
5. **Installing Cloud Pak for Data**: https://www.ibm.com/docs/en/cloud-paks/cp-data/4.0?topic=installing-cloud-pak-data
6. **Operator life cycle manager**: https://olm.operatorframework.io/
7. **RHEL CoreOS**: https://docs.openshift.com/container-platform/4.6/architecture/architecture-rhcos.html
8. **CRI: Container Runtime Interface**: https://github.com/kubernetes/community/blob/master/contributors/devel/sig-node/container-runtime-interface.md
9. **CRI-O**: https://github.com/cri-o/cri-o
10. **Kubernetes cluster architecture**: https://kubernetes.io/docs/concepts/architecture/
11. **Kubernetes components**: https://kubernetes.io/docs/concepts/overview/components/

12. **Kubernetes Deployments**: https://kubernetes.io/docs/concepts/
workloads/controllers/deployment/

13. **Kubernetes StatefulSets**: https://kubernetes.io/docs/concepts/
workloads/controllers/statefulset/

14. **Kubernetes services**: https://kubernetes.io/docs/concepts/
services-networking/service/

15. **Networking**: https://kubernetes.io/docs/concepts/services-
networking/

16. **Configuring ingress**: https://docs.openshift.com/container-
platform/4.7/networking/configuring_ingress_cluster_
traffic/overview-traffic.html

17. **Exposing a route**: https://docs.openshift.com/
container-platform/4.7/networking/configuring_
ingress_cluster_traffic/configuring-ingress-
cluster-traffic-ingress-controller.html#nw-exposing-
service_configuring-ingress-cluster-traffic-ingress-
controller

18. **Secured routes**: https://docs.openshift.com/container-
platform/4.7/networking/routes/secured-routes.html

19. **Using a custom TLS certificate**: https://www.ibm.com/docs/en/cloud-
paks/cp-data/4.0?topic=client-using-custom-tls-certificate

20. **Dynamic storage provisioning**: https://kubernetes.io/docs/
concepts/storage/dynamic-provisioning/

21. **Operand Deployment Lifecycle Manager**: https://github.com/IBM/
operand-deployment-lifecycle-manager

22. **Namespace scope operator**: https://github.com/IBM/ibm-namespace-
scope-operator

23. **Container Application Software for Enterprises (CASE) specification**:
https://github.com/IBM/case

24. **cpd-cli**: https://www.ibm.com/docs/en/cloud-paks/
cp-data/4.0?topic=administering-cpd-cli-command-reference

25. **Service labels and annotations**: https://www.ibm.com/docs/en/cloud-
paks/cp-data/4.0?topic=alerting-monitoring-objects-by-
using-labels-annotations

26. **Cloud Pak for Data scheduler**: https://www.ibm.com/docs/en/cloud-
paks/cp-data/4.0?topic=service-installing-scheduling

27. **Alerting**: `https://www.ibm.com/docs/en/cloud-paks/cp-data/4.0?topic=platform-monitoring-alerting`

28. **Custom monitors**: `https://www.ibm.com/docs/en/cloud-paks/cp-data/4.0?topic=alerting-custom-monitors`

29. **Compute resources for containers in Kubernetes**: `https://kubernetes.io/docs/concepts/configuration/manage-resources-containers/`

30. **Service operator subscriptions**: `https://www.ibm.com/docs/en/cloud-paks/cp-data/4.0?topic=tasks-creating-operator-subscriptions#preinstall-operator-subscriptions__svc-subcriptions`

31. **Kubernetes namespace resource quotas**: `https://kubernetes.io/docs/tasks/administer-cluster/manage-resources/quota-memory-cpu-namespace/`

32. **Storage volumes**: `https://www.ibm.com/docs/en/cloud-paks/cp-data/4.0?topic=platform-managing-storage-volumes`

33. **Storage volumes API**: `https://www.ibm.com/docs/en/cloud-paks/cp-data/4.0?topic=resources-volumes-api`

34. **Kubernetes namespaces**: `https://kubernetes.io/docs/concepts/overview/working-with-objects/namespaces/`

35. **OpenShift projects**: `https://docs.openshift.com/container-platform/4.7/applications/projects/working-with-projects.html`

36. **Operator subscription InstallPlans**: `https://www.ibm.com/docs/en/cloud-paks/cp-data/4.0?topic=tasks-creating-operator-subscriptions#preinstall-operator-subscriptions__install-plan`

37. **Scaling services**: `https://www.ibm.com/docs/en/cloud-paks/cp-data/4.0?topic=cluster-scaling-services`

38. **Backup and restore**: `https://www.ibm.com/docs/en/cloud-paks/cp-data/4.0?topic=cluster-backing-up-restoring-your-project`

39. **Offline backup via filesystem copy**: `https://www.ibm.com/docs/en/cloud-paks/cp-data/4.0?topic=bu-backing-up-file-system-local-repository-object-store`

40. **Backing up with a snapshot**: `https://www.ibm.com/docs/en/cloud-paks/cp-data/4.0?topic=up-backing-file-system-portworx`

10
Security and Compliance

Cloud Pak for Data is an offering for enterprises that have stringent requirements. Cloud Pak for Data as a multi-user platform is also expected to provide governance in the implementation of separation of duties, auditing, and other compliance requirements.

The **Cloud Pak for Data security white paper** [1] describes many security aspects, starting with the development of software to deployment and operations.

For a containerized platform offering such as Cloud Pak for Data, where multiple services operate in a co-located manner and different user personas are expected to access the system, there are strict guidelines on how these services are developed and delivered, as well as how these services are to be managed and monitored in enterprise data centers. *Chapter 9, Technical Overview, Management and Administration* introduced the core concepts of the architecture stack, including how clusters can even be shared and operated securely for different tenants with sufficient isolation.

In this chapter, we will begin by exploring how IBM ensures security during the development of Cloud Pak for Data and how the stack ensures security from the ground up. Security Administrators will gain insight into specific operational considerations and techniques for safely operating Cloud Pak for Data on behalf of their users, as well as, learning of mechanisms to satisfy regulatory compliance requirements.

In this chapter, we will be exploring these different aspects:

- Secure engineering – how Cloud Pak for Data services are developed securely
- Secure operations in a shared environment
- User access control and authorizations
- Meeting compliance requirements

Technical requirements

For a deeper understanding of the topics described in this chapter, you are expected to have some familiarity with the following technologies and concepts:

- The Linux operating system and its security primitives
- Cloud-native approaches to developing software services
- Virtualization, containerization, and Docker technologies
- Kubernetes (Red Hat OpenShift Container Platform is highly recommended)
- Storage provisioning and filesystems
- Security hardening of web applications, Linux hosts, and containers
- The Kubernetes operator pattern (highly recommended)
- **The kubectl and oc** command-line utilities, as well as the YAML and JSON formats needed to work with Kubernetes objects
- An existing Cloud Pak for Data installation on an OpenShift Kubernetes cluster, with administration authority

The sections will also provide links to key external reference material, in context, to help you understand a specific concept in greater detail or to gain some fundamental background on the topic.

Security and Privacy by Design

The objective of **Security and Privacy by Design** (**SPbD**) is to ensure that the best practices of secure engineering are followed during the development of any offering and to implement processes for the continuous assessment of the security posture of that product. Release engineering processes and timely remediation of any security incidents or discovered vulnerabilities via **Product Security Incident Response Team** (**PSIRT**) mandates are also critical to reduce the risk of exposures and to protect against malicious actors who may compromise the system. IBM's secure development practices are described in detail in the Redbook [2]. IBM also has a formal process to track and respond to any vulnerabilities via the Product Security Incident and Response team [3].

Development practices

IBM requires all products to be developed and evaluated using strict secure engineering practices, and the services delivered for Cloud Pak for Data are no exception. With such offerings also being operated in an as-a-service fashion, accessible to the general public, security is of paramount importance right from the design of the software.

The security practices and reports are independently reviewed by the IBM hybrid cloud **Business Information Security Office (BISO)** prior to any release of the software.

This section describes some of the important practices used to develop security-hardened services in Cloud Pak for Data.

Vulnerability detection

A key initiative is to detect possible security weak points or potential areas that a malicious operator could leverage to introduce man-in-the-middle attacks, access sensitive data, or inject malware. Given that quite a bit of interaction with Cloud Pak for Data is through a web browser or HTTP-based clients in general, care is taken to ensure protection from both outside and in-cluster influences:

- **Threat modeling**: This is a process to analyze where the system is most vulnerable to attack, by building out inter-service and intra-service detailed data flow diagrams, understanding potential attack vectors, as well as evaluating whether existing protection is deemed sufficient. The output of the threat modeling exercise is used to drive the development of additional security constructs or to refactor service interactions to mitigate potential threats.

- **Security quality assurance and code reviews**: Additional tests are developed by a team focused on security and operating independently from typical functional development and quality assurance squads. This effort also involves validation of cross-service/cross-product integrations and a thorough evaluation of general system reliability using **Chaos Monkey** style testing approaches (that is, mechanisms to generate random failures in the system to validate resiliency).

  This allows for a better objective evaluation of a proposed software delivery against required security guidelines. As part of this initiative, this team also conducts independent code reviews and approval of any mitigations proposed by the service development teams.

- **Code scans**: Static code scans and web application scans are mandatory, and a key requirement is to identify and mitigate threats as described by the **Open Web Application Security Project** [4] initiative. IBM Security AppScan for static source code is typically run as part of daily builds of the software and dynamic web application scanning is performed side by side with functional verification tests.

- **Penetration testing**: Deep testing to mimic cyber threats to identify security flaws, vulnerabilities, and unreliable systems, and in general, multiple ethical hacking techniques are required to be performed as part of every major release. This is required to be performed by a specialist organization or third-party company for independent evaluation of the deployed software.

  Any findings are reviewed independently by the IBM BISO organization and tracked for immediate resolutions, even via software patches or by publishing mitigation guidelines.

In this section, we have looked at the secure engineering processes that are employed in the development of Cloud Pak for Data software. In the next section, we will look at how this software is released for customers to access in a secure manner from a trusted source.

Delivering security assured container images

Each Cloud Pak for Data service is delivered as container images and loaded into a registry. Such images are composed of various layers, with each layer bringing in multiple files and operating system packages. To mitigate risks associated with deploying such images, IBM processes require Cloud Pak for Data services to use a number of techniques prior to a release.

Vulnerability scanners

Common Vulnerabilities and Exposures (**CVEs**) are publicly disclosed security issues found in operating system libraries and open source packages. A fundamental risk reduction mechanism is to frequently scan all images and quickly mitigate any issues found. The IBM Vulnerability Advisor, a sophisticated service offered in the IBM Cloud, scans images for known **CVEs**, insecure operating system settings, exposed sensitive information in configuration files, and misconfigured applications.

Certified base images

Red Hat makes available secure **Enterprise Linux** (**RHEL**) base images [5] for offerings such as Cloud Pak for Data to leverage.

The Red Hat **Universal Base Images** (**UBIs**) provide for a supported set of operating systems and open source packages. These images also contain a much-reduced set of packages than the typical RHEL distribution and thus expose a smaller attack surface. Red Hat continuously tracks security and reliability issues on all packages and provides new versions of these base images.

By using such certified base images, Cloud Pak for Data services are able to provide a trusted and secure foundation for their functionality.

Signing

Code signing and container image signing enable enterprises to validate the IBM provenance of the Cloud Pak for Data software and ensure the integrity of the containers that run within their Kubernetes clusters. This practice reduces the risk of any malicious injection into the released container images.

The IBM Container Registry (ICR)

Container images for Cloud Pak for Data are hosted in ICR. This enables the continuous delivery of images, including security fixes and monitoring these images for vulnerabilities via online scanners, and so on.

This registry is secure, not open to the public, and access is available to only customers who have entitlement to these services. These customers will be able to pull these images with their own keys [6] and with a registered IBM ID.

Operator Lifecycle Manager (OLM)

OpenShift v4 includes OLM [7] out of the box and a prescriptive method to introduce operators, such as from Cloud Pak for Data services, into the cluster as well as to enable the continuous delivery and update of these operators. This makes it easier for cluster administrators to securely manage all operators and grant access to specific operators to individual projects within that cluster.

Supporting air-gapped clusters

Air-gapped data centers permit neither inbound nor outbound internet access. This makes it complicated for OpenShift clusters to directly pull container images from IBM's registry. While in some cases, IP whitelisting may be possible to allow such pulls, it may not work for many enterprises. Besides, there could be latency or network disruptions that impact the day-to-day operation of services in the OpenShift Kubernetes cluster.

Hence, it is expected that all OpenShift clusters are able to pull images from a designated *Enterprise* container registry in their own private network. It is expected that pull secrets are made available to the service accounts used by Cloud Pak for Data.

The role of a bastion node

Bastion nodes are typically machines that are deployed outside of the enterprise's private network, with controlled access to repositories and registries. Such hosts facilitate access to container images, utilities, scripts, and other content from IBM and make them available to OpenShift clusters for provisioning.

One approach is to download all images as `tar.gz` files and transfer these files into the private network where they can be loaded into an appropriate container registry. However, a more convenient approach is to use the container registry to directly replicate container images from ICR into the enterprise's own registry, from such a bastion node. This is made possible because the Bastion node has access to the internet and to the enterprise's own data center or at least the target registry for the duration. A bastion node's role may extend beyond just the initial installation since this approach is quite useful for pulling or replicating newer container images, including those that patch any serious vulnerabilities.

In this section, we covered how IBM Cloud Pak for Data containerized software is made available for customers from a trusted location for customers to ensure its provenance, and how these images can be introduced into the customer's own registry. In the next section, we will look at what it means to operate Cloud Pak for Data securely.

Secure operations in a shared environment

In traditional systems, many applications share the same operating system, be it a virtual machine or bare metal. However, for security reasons, these applications are typically never granted access to the operating system or run as *root*. Since many programs, from different products or vendors, would be operating on the same machine, care is also taken to isolate each of these operating system processes from each other.

While it has been common to simply spin up a separate virtual machine for each application to completely isolate them, it was also expensive to operate and could possibly lead to a waste of compute resources. Containers have proven to be much less expensive in the long run, but there is a trade-off with regard to less isolation of workloads. In this section, we will look at how the stack enables security from the ground up, starting with the host operating system in the cluster, to OpenShift security constructs as well as how Cloud Pak for Data leverages Kubernetes access control primitives to enforce controls on all Services operating in that cluster.

Securing Kubernetes hosts

OpenShift and **Red Hat Enterprise Linux** (**RHEL**) CoreOS provide certain constructs and safeguards to ensure the secure operation of containerized workloads on the same set of Kubernetes compute nodes. RHEL CoreOS is also considered to be immutable for the most part, and the included CRI-O container engine has a smaller footprint and presents a reduced attack surface.

Controlling host access

By default, access to the host systems for containerized workloads on OpenShift is controlled to prevent vulnerabilities in such workloads from taking over the host.

Even for the installation of Cloud Pak for Data software, logging in to the OpenShift hosts is not required nor desired. None of the running containers are permitted to expose **Secure Shell** (**SSH**) style access either.

The RHEL CoreOS hosts provide specific configuration settings *[8]* out of the box to ensure security even when different workloads are running on the same host.

SELinux enforcement

RHEL provides a construct called **Security-Enhanced Linux** (**SELinux**) that provides a secure foundation in the operating system *[9]*. It provides for isolations between containers and **mandatory access control** (**MAC**) for every user, application, process, and file and thus dictates how individual processes can work with files and how processes can interact with each other.

OpenShift requires SELinux policy to be set to "enforcing" and thus forces security at the operating system level to be hardened. SELinux provides finely-grained policies that define how processes can interact with each other and access various system resources.

Security in OpenShift Container Platform

Service accounts are authorization mechanisms that allow for a particular component to connect to the Kubernetes API server and invoke specific actions as part of its day-to-day operations. Kubernetes roles are bound to such service accounts and dictate what exactly that service account is permitted to do.

OpenShift provides additional constructs on top of Kubernetes to harden security in a cluster over and above roles.

Security Context Constraints (SCCs)

The concept of **Security Context Constraints (SCCs)**[10] allows cluster administrators to decide how much privilege to associate with any specific workload. SCCs define the set of conditions a Kubernetes pod must adhere to for it to be scheduled on the platform. The permissions granted, especially to access host resources such as filesystems or networks, or even which user IDs or group IDs a Pod's processes can run as, are governed by the associated SCCs. For example, the *privileged* SCC allows *all* privileges and access to host directories, even run as root, while the *restricted* SCC denies *all* of that.

By default, the *restricted* SCC is used for all Pods, while anything over *restricted* is considered an exception that a cluster administrator needs to authorize by creating a binding of that SCC to a service account.

With *restricted*, running processes as root is disabled and Linux capabilities are dropped, such as *KILL, MKNOD, SETUID, SETGID*, and so on. This SCC prevents running privileged containers as well. Host paths, host network access, and **Inter-Process Communication** (IPC) are denied. This is enforced by OpenShift when Pods are started up and any attempt by processes within the Pod to violate this policy results in the Pod being shut down with appropriate errors.

Cloud Pak for Data **v3** introduced additional SCCs that granted incremental additional capabilities over *restricted* for some of its Pods without needing to use far more open SCCs such as *privileged* or *anyuid*:

- **cpd-user-scc**:- This custom SCC added a fixed UID range (1000320900 to 1000361000), which allows all Pods that were associated with this SCC to be run with these user IDs. This allowed different Pods to be deterministically run with well-known user IDs and helped with the management of file ownership. Note: running as root is still prohibited. This policy also removed the need to assign an FSGroup policy to every Pod, which could have adverse effects in setting filesystem permissions for shared volumes.

  The FSGroup policy is needed to work around a problem where Kubernetes recursively resets file system permissions on mounted volumes. While this has a serious performance impact for large volumes, even causing timeouts and failures when pods start up, a more serious problem is associated with file permissions getting reset and perhaps made accessible to more Pods and users than originally intended.

  A future capability in Kubernetes (*fsGroupChangePolicy*) is expected to help workloads avoid this problem. *[12]*

- **cpd-zensys-scc:** This custom SCC is identical to `cpd-user-scc`, except that it also allows SETUID/SETGID privileges. It also designates one UID, 1000321000, as a *System* user. The primary reason for this SCC is to enable system Jobs and containers to initialize persistent volumes with the right file system ownership and permissions to allow individual content in shared volumes to be safely read or modified by other Pods.

> **Note**
>
> With Cloud Pak for Data **v4**, and since **OpenShift Container Platform v3.11** is reaching end-of-support, and with OpenShift Container Platform v4.6 providing additional security constructs needed for Cloud Pak for Data, these custom SCCs are no longer created. However, some services such as DB2 and Watson Knowledge Catalog continue to require additional custom SCCs in v4 as well. These are described in the service documentation [37].

Namespace scoping and service account privileges

Cloud Pak for Data services are deployed into specific OpenShift project namespaces and their service accounts are scoped to only operate within those namespaces. These service accounts are associated with various Kubernetes Deployments running in those namespaces.

OpenShift assigns a unique **Multi-Category Security** (**MCS**) label in SELinux to each OpenShift project. This ensures that Pods from one namespace cannot access files created by Pods in another namespace or by host processes with the same UID. *[11]*.

RBAC and the least privilege principle

The concept of roles in Kubernetes helps Cloud Pak for Data employ the principle of least privilege to grant each Pod only as much authorization for actions as needed and nothing more. This is done by introducing the following roles into the namespace where Cloud Pak for Data is deployed:

Role	Description
cpd-admin-role	Allows creation/updates/deletes of Kubernetes deployments, secrets, configmaps, run Jobs, and more, but only within that specific namespace.
cpd-viewer-role	Allows Kubernetes API calls (GETs) but cannot perform any creates or deletes.

Figure 10.1 – Kubernetes roles used by Cloud Pak for Data service accounts

Specific service accounts are introduced into the namespace, which are then *bound* to these roles. That defines the scope of privileges each account is granted.

> **Note**
>
> With Cloud Pak for Data v4, since custom SCCs are no longer required, a different set of service accounts, with the prefix *zen-* were introduced to ensure that any existing SCCs are no longer picked, even accidentally.

Kubernetes pods run with the authority of a service account. Apart from being only permitted to work within the confines of a namespace, the Cloud Pak for Data Service accounts are designed to assign only the least privileges needed for the specific actions being performed. The following table represents the key namespace scoped service accounts used in Cloud Pak for Data with v3.5 and v4 with details about what they are used for and what they are permitted to do.

Account [v3 and v3.5]	Description
cpd-admin-sa	Bound to the cpd-admin-role and is associated with a system user and cpd-zensys-scc. This account is used for preparing persistent volume mounts, setting file permissions, and more.
cpd-editor-sa	Bound to the cpd-admin-role also but is associated with the cpd-user-scc, and thus has reduced privileges.
cpd-viewer-sa	Bound to cpd-viewer-role and cpd-user-scc. This account can only view Kubernetes artifacts within that namespace.
cpd-norbac-sa	Bound to cpd-noperm-scc but is not associated with any Kubernetes role thus has no API access. This account is similar to the out-of-the-box default service account. It is used by services or components that should not be granted any RBAC privileges.
default	Automatically created in every OpenShift project, bound to the restricted SCC, and is not granted any RBAC privileges; that is, no roles are bound. This default service account will be used for end user workloads such as notebooks and Python Jobs and will not be allowed to perform any kind of actions inside the namespace.

Figure 10.2 – Namespace scoped service accounts used in Cloud Pak for Data – with v3 and v3.5

Following is the Namespace scoped service accounts used in Cloud Pak for Data – with **v4**:

Account [v4]	Description
zen-admin-sa	Bound to the cpd-admin-role and, by default, associated with the OpenShift "restricted" SCC.
zen-editor-sa	Bound to the cpd-admin-role and associated with the OpenShift "restricted" SCC.
zen-viewer-sa	Bound to the cpd-viewer-role and, by default, with the OpenShift "restricted" SCC. This account can only view Kubernetes artifacts within that namespace.
zen-norbac-sa	Associated, by default, with the OpenShift "restricted" SCC and is not associated with any Kubernetes role, thus has no API access. This account is similar to the out-of-the-box default service account. It is used by services or components that should not be granted any RBAC privileges.
default	Automatically created in every OpenShift project, bound to the restricted SCC, and is not granted any RBAC privileges; that is, no roles are bound. This default service account will be used for end user workloads such as notebooks and Python Jobs and will not be allowed to perform any kind of actions inside the namespace.

Figure 10.3 – Namespace scoped service accounts used in Cloud Pak for Data – with v3.5 and v4

Hence, system Pods and Jobs typically use **cpd-admin-sa** to perform higher-privileged operations, whereas regular user Pods and Jobs (such as data science environments and refinery Jobs) would be associated with **default** or **cpd-norbac-sa** to designate that they are not granted any privileges whatsoever.

A typical Cloud Pak for Data Pod spec, as shown in the following code snippet, includes `securityContext` directives that indicate which Linux capabilities to add or drop, as well as which user ID to run as, and even identify explicitly that the Pod must only be run with a non-root UID. The service account is also identified, which in turn grants the Pod specific RBAC in terms of working with Kubernetes resources and APIs:

```
securityContext:
        capabilities:
            drop:
              - ALL
```

```
        runAsNonRoot: true
  serviceAccount: cpd-editor-sa
  serviceAccountName: cpd-editor-sa
```

Workload notification and reliability assurance

Cloud Pak for Data also follows a specific convention where every container in its namespace is clearly tagged with a specific set of labels and annotations. This allows the clear identification of each Pod or Job running in the cluster, which proves crucial in identifying any rogue or compromised workloads.

The following code block shows an example of such a convention:

```
  annotations:
   cloudpakName: "IBM Cloud Pak for Data
   productID: "eb9998dcc5d24e3eb5b6fb488f750fe2"
   productName: "IBM Cloud Pak for Data Control Plane"
  labels:
          app=0020-zen-base
          component=zen-core-api
           icpdsupport/addOnId=zen-lite
           icpdsupport/app=framework
           release=0020-core
```

With dynamic workloads and sharing of compute, "noisy neighbor" situations can occur, where workloads grow to consume all available compute resources leading to an adverse impact on the stability of the cluster. To avoid such problems, every Pod and Job spun up in the namespace is also associated with resource requests and limits.

Cloud Pak for Data provides mechanisms to scale up and scale out compute, and by explicitly setting such compute specifications, it ensures that each Pod or Job sticks to its assigned compute resources and supports the reliable operation of a balanced cluster.

The following code block shows a Pod spec including a resource specification:

```
  resources:
          limits:
              cpu: 500m
              memory: 1Gi
```

```
requests:
    cpu: 100m
    memory: 256Mi
```

Additional considerations

In the previous sections, we looked at how Cloud Pak for Data workloads can be uniquely identified, how the least privilege principle is used to grant RBAC to each microservice, and even how resources consumed by each micro-service can be capped. In this section, we will cover aspects of security that apply across *multiple* microservices or even the entire Cloud Pak for Data installation.

We will explore what it means to do the following:

- Ensure security for data in motion between two microservices

- Secure data at rest

- Understand the relevance of traditional anti-virus software

Encryption in motion and securing entry points

Cloud Pak for Data exposes one HTTPS port as the primary access point for the browser-based user experience and for API-based access. This port is exposed as an OpenShift route. All communication inside the namespace and inbound is encrypted using SSL, with only **TLS 1.2** highly secure cryptographic ciphers being enabled.

The installation of Cloud Pak for Data in an OpenShift namespace causes the generation of a self-signed TLS certificate. By default, this certificate is untrusted by all HTTPS clients and should be replaced with the enterprise's own signed TLS certificate *[13]*.

All services in Cloud Pak for Data are required to support TLS-based encryption for service-to-service communication and for communicating with external systems. For example, DB2 exposes TLS over ODBC and JDBC for client applications, even from outside the cluster to connect securely and ensure data in transit is also encrypted. Connecting to external customer-identified databases and data lakes in services such as Data Virtualization, Watson Studio can also be configured to be over TLS.

Note that such in-transit TLS-based encryption in Cloud Pak for Data services is over and above anything that may be configured in the infrastructure. For example, it is possible that Kubernetes inter-node communication networks are also encrypted as part of a cloud infrastructure or VPN techniques.

Encryption at rest

OpenShift Container Platform supports the encryption of the etcd database *[14]* that stores secrets and other sensitive Kubernetes objects that are critical for day-to-day operations.

By default, only Pods within the same namespace are granted access to mount persistent volume claims. However, it is also important to encrypt data stored in these Persistent Volumes as well to avoid not just malicious actors, but also admins or operators themselves from getting access to sensitive data. Typical regulatory requirements, such as compliance to the separation of duties, require that client data is kept private and out of reach of operators. Encryption goes a long way in ensuring security and compliance in the platform.

Cloud Pak for Data supports multiple storage providers on OpenShift, each with its own techniques to encrypt volumes. Some examples are listed as follows:

- **Portworx** *[15]* provides the ability to encrypt all volumes with a cluster-wide secret passphrase or each individual volume with its own secret passphrase. Portworx also integrates with typical cloud key management systems to support the bring-your-own-keys requirement.

- **Red Hat OpenShift Container Storage v4.6** and later *[16]* supports encryption of the whole storage cluster with a per-device encryption key stored as a Kubernetes Secret.

- **IBM Cloud File Storage** *[17]* with Endurance or Performance options support for encryption-at-rest by default and encryption keys managed in-house with the **Key Management Interoperability Protocol** (**KMIP**).

- **Network file system** (**NFS**) volume encryption depends on the underlying storage provider. The standard **Linux Unified Key Setup-on-disk-format** (**LUKS**) *[18]* can be leveraged as a general disk encryption technique.

While *raw* block or filesystem-based data-at-rest encryption is essentially beyond the scope of Cloud Pak for Data and is heavily reliant on storage infrastructure, it must be noted here that Cloud Pak for Data, by itself, ensures encryption for *specific* sensitive data. Examples of such data are user credentials for database connections or service keys, and these are encrypted *natively* by the appropriate service in its own repository, without any dependency on the underlying storage technology.

Anti-virus software

The use of traditional anti-virus software can have an adverse effect on OpenShift. For example, such software may lock critical files or consume CPU in an ad hoc fashion in the operating system, stepping around OpenShift's careful management of resources.

It is best *not* to think of these Kubernetes nodes as general-purpose Linux systems, but rather as specialized machines secured and managed by OpenShift as a cluster. OpenShift provides security and operations for the *entire stack* from the operating system to containers to individual workloads. Traditional anti-virus software cannot cover the entire stack and may interfere with the operations or reliability of the cluster.

Red Hat and OpenShift provide additional security hardening to reduce the need for anti-virus software in the first place *[19]*. Note that traditional scanners are typically used to locate viruses in shared file systems (such as those served by Samba) and mounted on Windows clients. But, with the prescriptive approach for operating Kubernetes within OpenShift Container Platform and on RHEL CoreOS, such arbitrary file serving should *not* be set up on these Kubernetes nodes in the first place.

In this section, we explored the security constructs that are available in the Operating System and OpenShift. We also looked at how Cloud Pak for Data, with the judicious use of Kubernetes RBAC can ensure the least privileges for each service account and uses a convention of labels and annotations to tag all workloads. It also constrains the assignment of compute resources to ensure secure and reliable operations. In the following section, we will look at what controls are available for Cloud Pak for Data administrators to configure authentication and ensure the right level of authorization for each user.

User access and authorizations

Cloud Pak for Data is a platform into itself, is a multi-user system, and thus also provides access control for users. It also integrates with Enterprise LDAP or Active Directory and other identity providers to ensure secure access to its services. Auditing is also supported for tracking access to the Cloud Pak for Data platform as well.

As with any multi-user system, Cloud Pak for Data identifies roles and permissions that are granted to specific users or groups of users. A suitably privileged user is tasked with the responsibility of ensuring that only authorized users can access the services and content hosted by Cloud Pak for Data.

Authentication

While Cloud Pak for Data includes a simple out-of-the-box mechanism to identify users and grant them access, this is only for initial runs and is not meant for secure production use, and will fail to meet any compliance regulation requirements. For example, password policies or multi-factor authentication cannot be supported with this default mechanism. Instead, the expectation is that the enterprise sets up connectivity to an identity provider.

Identity providers and SSO

Cloud Pak for Data provides a user experience for configuring LDAP/Active Directory for authentication, as shown in the following figure:

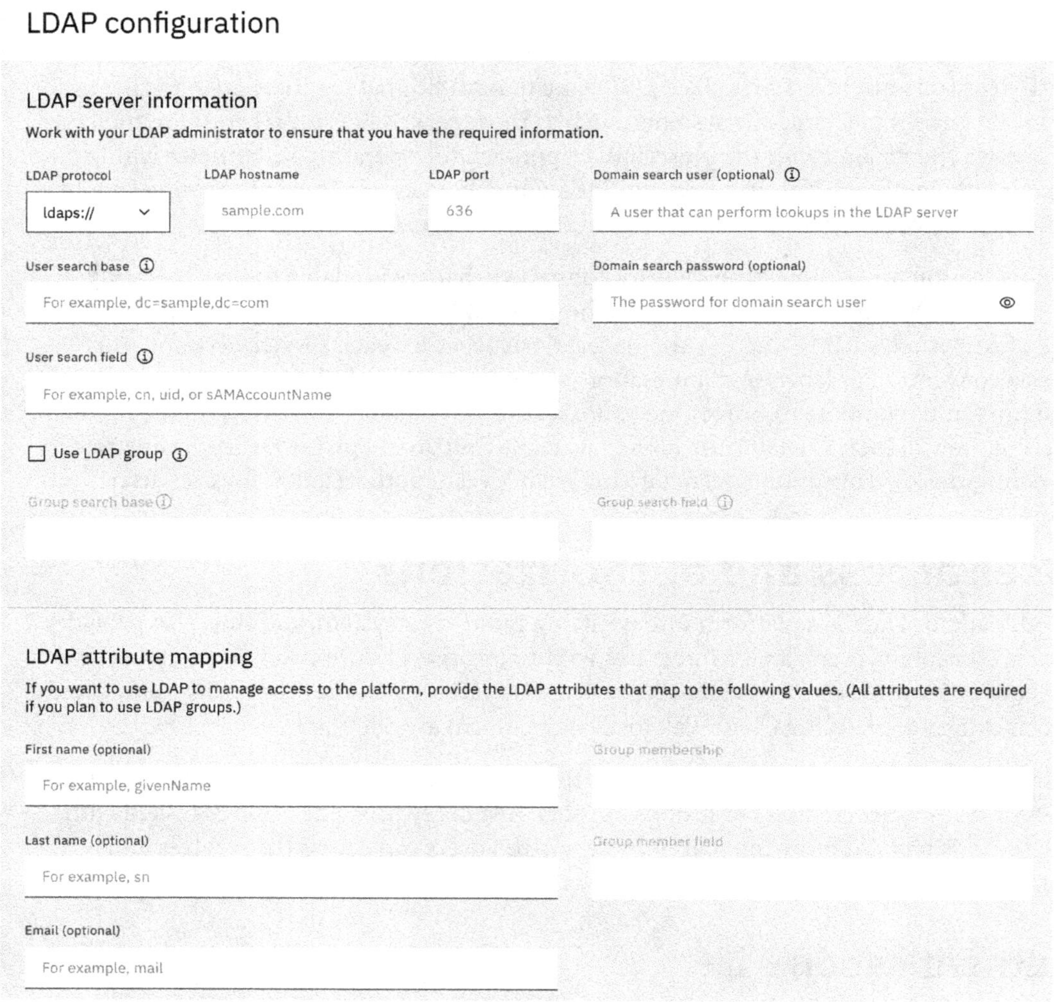

LDAP configuration

LDAP server information

Work with your LDAP administrator to ensure that you have the required information.

LDAP protocol	LDAP hostname	LDAP port	Domain search user (optional) ⓘ
ldaps:// ⌄	sample.com	636	A user that can perform lookups in the LDAP server

User search base ⓘ	Domain search password (optional)
For example, dc=sample,dc=com	The password for domain search user 👁

User search field ⓘ

For example, cn, uid, or sAMAccountName

☐ Use LDAP group ⓘ

Group search base ⓘ

Group search field ⓘ

LDAP attribute mapping

If you want to use LDAP to manage access to the platform, provide the LDAP attributes that map to the following values. (All attributes are required if you plan to use LDAP groups.)

First name (optional)	Group membership
For example, givenName	

Last name (optional)	Group member field
For example, sn	

Email (optional)

For example, mail

Figure 10.4 – LDAP setup for authentication

When LDAP is configured, authentication is delegated to the LDAP server for every sign-in to Cloud Pak for Data. Similarly, when **Security Assertion Markup Language (SAML)** Web **Single Sign-On (SSO)** *[20]* is set up, sign-in is redirected to the SAML log-in page instead.

Cloud Pak for Data can also be configured to cache a list of users and LDAP groups that were granted access to Cloud Pak for Data. This can be done to reduce querying LDAP for every lookup and avoid any adverse impact on the LDAP server and other enterprise services. Cloud Pak for Data provides an asynchronous Job that can be enabled and configured to run at a fixed interval to keep this cache consistent with each user's profile and group membership in LDAP.

Cloud Pak for Data also supports integrating [22] with IBM Cloud Platform Common Services IAM [21] as a mechanism to work with other identity providers and to support SSO. This is particularly useful when there are multiple Cloud Paks installed on the same OpenShift cluster. Once Cloud Pak for Data administrator privileges have been granted to a user who is authenticated by an identity provider or LDAP/**Active Directory** (**AD**), it is recommended that such an admin disable or remove all users from the internal out-of-the-box user repository, including the default admin user [23]. This is critical from an audit and compliance perspective as well, since all users would need to be authenticated by the prescribed mechanism, rather than via back doors that may exist.

Session management

Cloud Pak for Data relies on the concept of a bearer token (JSON Web Tokens) to identify a user's session. This token expires based on the policies defined in the Cloud Pak for Data configuration. For interactive usage, in the Cloud Pak for Data user experience in web browsers, the user's token is short-lived but is continuously renewed as long as the user is actively using the experience. A Cloud Pak for Data administrator can configure [24] the token expiry time and the web session token refresh period in accordance with enterprise security and compliance requirements. The web session token refresh period helps keep the user's session active. If a user is idle in their web browser for more than the configured length of time, then the user is automatically logged out of the web client.

There are two configuration parameters that govern user sessions:

- **Token expiry_time** – The length of time until a user's session expires.

  A user who is idle for this duration will find their browser session automatically logged out.
- **Token_refresh period** – The length of time that a user has to refresh their session.

This defines the longest the user can stay logged in. Once their session has reached this time limit, their tokens will no longer be refreshed – even if they are active all along – and will force them to log out.

Authorization

With permissions and roles, an enterprise can define their own personas and grant them specific access to different Cloud Pak for Data services.

A role is an entity that is used to define what collection of permissions (privileges) are granted to a particular user persona.

Cloud Pak for Data services introduce one or more out-of-the-box roles to get started with and an experience to help manage roles. Enterprises can use such roles to adhere to *separation of duties* policies needed for compliance reasons. For example, administrators may be denied access to data (which is the default for database instances they did not provision in the first place), and also to data catalogs, by revoking specific permissions.

Granular privileges can also be granted, for example, only granting a specific set of users the ability to manage user access and to manage group access, but they need not be granted access to other administrative functions.

The administrator typically configures roles using the user management experience, as shown in the following figure:

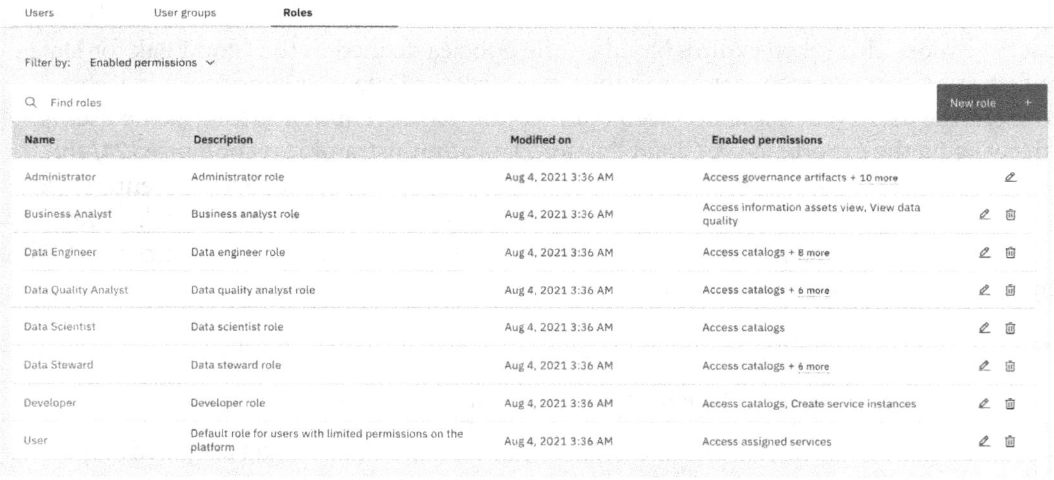

Name	Description	Modified on	Enabled permissions		
Administrator	Administrator role	Aug 4, 2021 3:36 AM	Access governance artifacts + 10 more	✎	
Business Analyst	Business analyst role	Aug 4, 2021 3:36 AM	Access information assets view, View data quality	✎	🗑
Data Engineer	Data engineer role	Aug 4, 2021 3:36 AM	Access catalogs + 8 more	✎	🗑
Data Quality Analyst	Data quality analyst role	Aug 4, 2021 3:36 AM	Access catalogs + 6 more	✎	🗑
Data Scientist	Data scientist role	Aug 4, 2021 3:36 AM	Access catalogs	✎	🗑
Data Steward	Data steward role	Aug 4, 2021 3:36 AM	Access catalogs + 6 more	✎	🗑
Developer	Developer role	Aug 4, 2021 3:36 AM	Access catalogs, Create service instances	✎	🗑
User	Default role for users with limited permissions on the platform	Aug 4, 2021 3:36 AM	Access assigned services	✎	🗑

Figure 10.5 – Managing roles experience

The following tables show examples of permissions assigned to the different out-of-the-box roles [38] in Cloud Pak for Data, as well as the associated service offerings:

- **Administrator** role: **Administrator** is a pre-defined role in Cloud Pak for Data and includes a specific set of permissions as listed in the following table. The set of permissions present depends on which services have been installed. For example, if Watson Knowledge Catalog is not present, then the **Access governance artifacts** permissions will not be available.

Permission	Category	Services that contribute the permission
Access governance artifacts	Governance artifacts	Watson™ Knowledge Catalog
Administer platform	Platform administration	Cloud Pak for Data control plane
Create deployment spaces	Deployments	− Watson Knowledge Catalog − Watson Studio
Create projects	Projects	− Watson Knowledge Catalog − Watson Studio
Create service instances	Service instances	Cloud Pak for Data control plane
Manage asset discovery	Data curation	Watson Knowledge Catalog
Manage catalogs	Catalogs	− Watson Knowledge Catalog − Watson Studio
Manage data protection rules	Governance artifacts	Watson Knowledge Catalog
Manage data quality	Data curation	Watson Knowledge Catalog
Manage data transformation	Data integration	DataStage®
Manage governance categories	Governance artifacts	Watson Knowledge Catalog
Manage information assets	Catalogs	Watson Knowledge Catalog
Manage metadata	Data curation	Watson Knowledge Catalog
Manage workflows	Workflows	Watson Knowledge Catalog

Figure 10.6 – Permissions granted to the Administrator role

- **Data Scientist** role: This role designates the user persona typically involved with advanced analytics and machine learning. They reference data from catalogs and collaborate with other users within projects. A small set of permissions is typically sufficient for such users, as listed in the following table. This also grants them the permission to work with deployments such as machine learning models.

Permission	Category	Services that contribute the permission
Access catalogs	Catalogs	– Watson Knowledge Catalog – Watson Studio
Create deployment spaces	Deployments	– Watson Knowledge Catalog – Watson Studio
Create projects	Projects	– Watson Knowledge Catalog – Watson Studio

Figure 10.7 – Permissions granted to the Data Scientist role

- **User** role: The **User** role is special as it typically indicates the least set of privileges assigned to a certain type of guest user. Assigning the **User** role to somebody just implies that they will be able to sign in and view any content that does not require any additional authorizations. Assigning the **User** role also serves to introduce new users to the platform, which enables them to be subsequently added (as individuals) into user groups or assigned access to projects, catalogs, or even databases.

However, when certain services, such as Watson Knowledge Catalog or Watson Studio are installed, by default, they grant additional permissions to this **User** role. This enables these users to at least work privately within the confines of their own projects or deployment spaces. Of course, just as with any pre-defined role, administrators can choose to alter this role and remove these permissions or introduce additional ones to suit their definition of a "least" privileged role.

Permission	Category	Services that contribute the permission
Create deployment spaces	Deployments	– Watson Knowledge Catalog – Watson Studio
Create projects	Projects	– Watson Knowledge Catalog – Watson Studio

Figure 10.8 – Permissions granted to the User role

The default user (admin) is automatically assigned all of the following roles out of the box, but this can be changed as needed:

- **Administrator**

- **Business Analyst**

- **Data Engineer**

- **Data Quality Analyst**

- **Data Scientist**

- **Data Steward**

- **Developer**

An enterprise can customize this by defining its own roles as well [25]. The admin can edit the out-of-the-box roles or create new roles if the default set of permissions doesn't align with business needs.

As part of a role's definition [26], specific permissions can be selected and associated with that role.

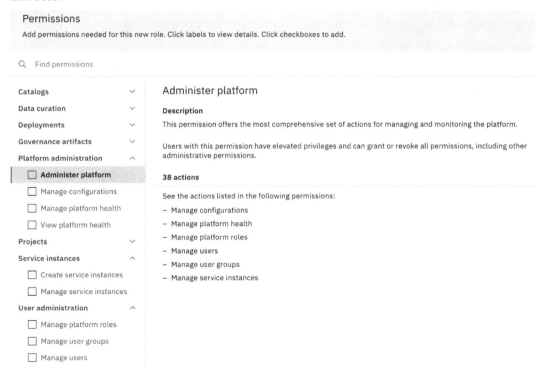

Figure 10.9 – Example of associating permissions with a new role

In this section, we discussed the semantics of permissions and how roles are defined or updated to suit the enterprise's requirements on access control and to identify suitable personas.

Roles are relevant to individual users as well as user groups, that is, roles can be assigned to Cloud Pak for Data User Groups, which implies that all members of that group will automatically inherit this role. In the following section, we will look more closely at how users are identified and groups are defined.

User management and groups

Cloud Pak for Data exposes APIs and a user experience for Admins to be able to manage authentication and role-based authorization for each user. The following figure provides an example of the user experience that helps the Admin grant and revoke access to users.

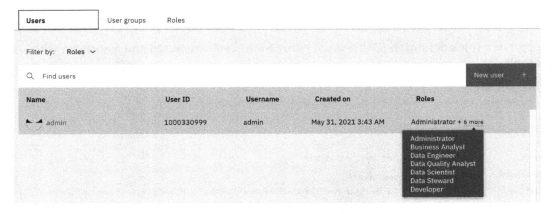

Figure 10.10 – Managing user access

An admin (or a user with the manage users permission) can authorize additional users to access Cloud Pak for Data.

For example, via the user experience, a new user could be introduced and assigned appropriate roles:

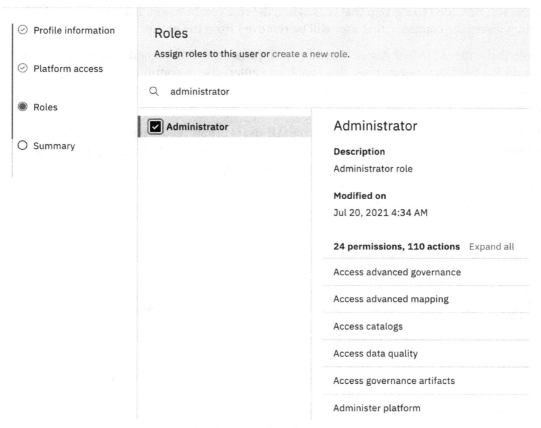

Figure 10.11 – Example of granting the Administrator role to a new user

As seen in the preceding figure, the **Administrator** granting such access can also review which permissions will be granted when this role assignment is completed. They could also choose to assign multiple roles to the same individual, granting them a union of the permissions from those roles.

Instead of granting permissions to individual users one at a time, Admins can also create user groups to simplify the process of managing large groups of users with similar access requirements. For example, if a set of users need the same combination of roles, then these users will be added to a group that is assigned the role combination. If a member of the group leaves the company, that user will be removed from the group.

Note that entire LDAP or Active Directory groups can also be assigned (mapped) to such Cloud Pak for Data user groups. This enables an enterprise to continue to leverage its existing organizational grouping, without needing to authorize every individual user one at a time.

By default, Cloud Pak for Data includes the **All-users** group. As the name suggests, all Cloud Pak for Data users are automatically included in this group. The group is used to give all platform users access to assets such as with **Connections**.

The user experience *[27],* shown in the following figure, enables the creation of such user groups and the assigning of roles to such groups:

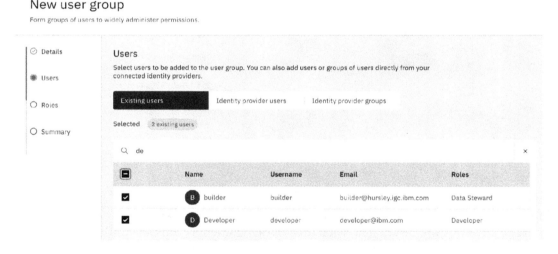

Figure 10.12 – Defining a new user group

To introduce members to such a user group, the Administrator also has the option to browse through any configured identity providers (such as LDAP) to identify individuals to add as well as *map* in existing LDAP groups. This provides quite a bit of flexibility as the Administrator is not limited by a fixed organizational structure in their Enterprise LDAP (say, as decided by HR) and can also, in ad hoc terms, enable sets of users even from different LDAP groups to collaborate.

The following figure illustrates how the experience enables browsing for existing LDAP groups to introduce as members of the Cloud Pak for Data user group.

New user group

Form groups of users to widely administer permissions.

⊘ Details	Users
● Users	Select users to be added to the user group. You can also add users or groups of users directly from your connected identity providers.
○ Roles	Existing users Identity provider users **Identity provider groups**
○ Summary	Selected None

Search for the identity provider groups you want to add to this user group.

🔍 A|

2 results returned

CN=**aaa** group,CN=Groups,DC=hursely,DC=igc,DC=ibm,DC=com

CN=**A**llowed RODC Password Replication Group,CN=Groups,DC=hursely,DC=igc,DC=ibm,DC=com

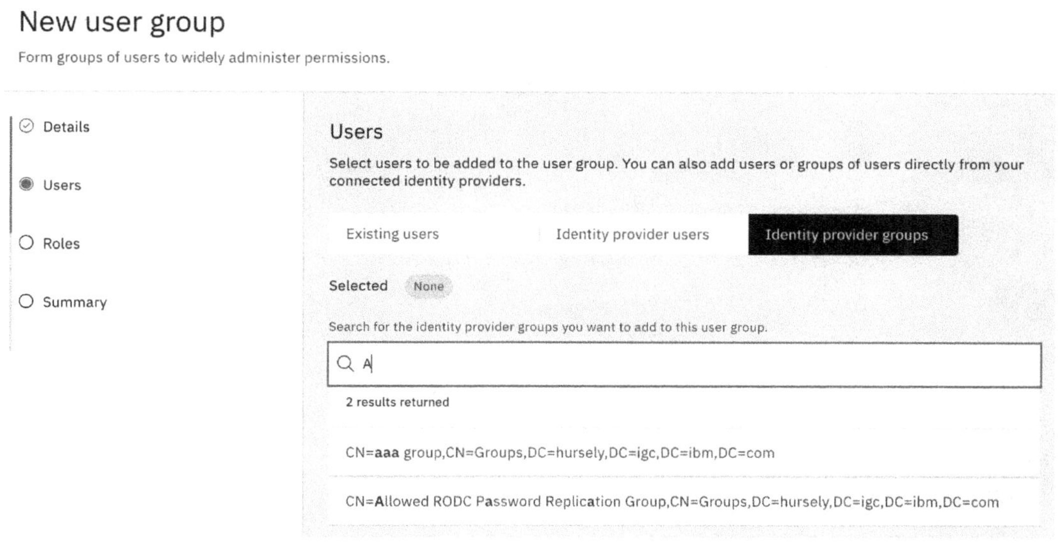

Figure 10.13 – Selecting existing LDAP groups to be assigned to a new User Group

As shown in the preceding figure, the experience enables the selection of existing fully qualified LDAP groups as members of the new Cloud Pak for Data User Group.

Once at least one individual user or LDAP group has been selected, the Administrator has the option to pick one or more roles to assign to this new User Group, as shown in the following figure:

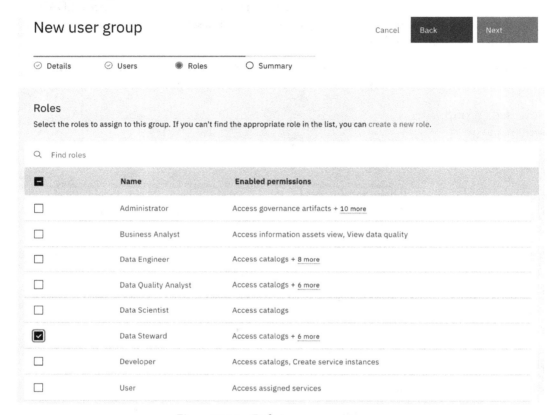

Figure 10.14 – Defining a new user group

In the preceding figure, you will notice that the **Data Steward** role has been assigned to the **Reviewers User Group**. This indicates that all members of this group will automatically be identified as data stewards and inherit the permissions that come with that role. The Administrator can also choose to change this role assignment at any time or introduce additional roles to this user group.

Note that the membership of a User Group and, by extension, the roles assigned to an individual, are *dynamic* when LDAP groups are used. This means that when a user is removed from an LDAP group or assigned to a different group, their subsequent sign-in will automatically ensure that their permissions reflect their current membership. An individual user may also be assigned to multiple User Groups, which means that their permissions represent a union of those inherited from their membership in all these groups.

In this section, we walked through how users and user groups are managed, and how Cloud Pak for Data integrates with enterprise authentication mechanisms, such as LDAP. Only after such a user is authenticated will they be able to access services or privileged resources in Cloud Pak for Data. In addition, there are also authorization checks. For example, users or user groups need to be granted at least a viewer role in resources such as projects or catalogs before being able to access them. Such users or groups could similarly be granted access to the **Data Virtualization Service** (**DVS**) instance as a *DV Engineer* or perhaps just as a *Database user* to an instance of a DB2 database. In summary, the User Management service supports a broad range of security requirements as it enables both authentication and flexible RBAC in the platform as well as authorization mechanisms scoped to individual resources or service instances.

Securing credentials

Cloud Pak for Data is a platform that provides several capabilities and services. It is expected that users and applications may need to connect from outside the cluster to specific API endpoints in a Cloud Pak for Data instance. Developers may also need to connect to external systems, such as databases and data lakes, as part of their day-to-day activities, for example, programmatically from notebooks or as part of Apache Spark Jobs. Such code is typically shared with other users, possibly even committed to source code repositories and, as such, it would not be prudent to have passwords and other sensitive content.

In this section, we will look at features in Cloud Pak for Data that provide the ability for such developers to avoid exposing their credentials in plain text.

API keys

Developers who invoke Cloud Pak for Data APIs from client programs can use API keys to avoid the need to persist their credentials within those clients.

Cloud Pak for Data v3.5 and later supports API keys for authentication and authorization. This allows client programs to authenticate to the Cloud Pak for Data platform or to a specific instance of a service.

Each user can generate a platform API key, which is of a broader scope and allows access to all services on the platform.

An instance API key is a particular service, and thus is only usable for that specific scope. The platform API key enables client scripts and applications to access everything that they would typically be able to access when they directly log in to the Cloud Pak for Data web client. An instance API key enables access only to the specific instance from which is it generated.

> **Note**
> The instance API key is not available for some services in the Cloud Pak for Data v3.5 release.

Cloud Pak for Data provides user interfaces and APIs *[28]* to generate, re-generate, or delete a platform API key or an instance API key.

The Secrets API and working with vaults

The Secrets API *[29]* in Cloud Pak for Data provides the ability to access sensitive credentials from configured vaults. This helps avoid the need to use plain-text credentials as part of developer-authored content such as notebooks or scripts. Using a vault, Enterprises can also control policies to rotate keys and credentials independently.

Since Cloud Pak for Data's Secrets API retrieves the contents of the vault only when required, it will get the latest password or key. This also prevents the need for users to update passwords in many places.

Cloud Pak for Data will continue to introduce integrations with different vault vendors and technologies as part of its roadmap.

Meeting compliance requirements

Regulatory compliance and enterprise policies require stringent monitoring of systems to ensure that they are not compromised. There are usually multiple standards to adhere to and different audit techniques that are required. Red Hat and IBM Cloud Pak for Data provide tools and mechanisms for security architects to design and configure the system as per the policies of their organization as well as to audit usage.

In this section, we will take a brief look at the capabilities that help enterprises plan for compliance.

Configuring the operating environment for compliance

For Cloud Pak for Data, Red Hat's Enterprise Linux for the host and OpenShift Container Platform represent the technological foundation. Hence, to ensure compliance, the foundation needs to be properly configured. A few key aspects of the technical stack that help in compliance readiness are listed as follows:

- RHEL supports a strong security and compliance posture such as setting up for the **Federal Information Processing Standard (FIPS)** publication 140-2. Several advanced techniques exist to secure RHEL hosts *[30],* including tools, scanners, and best practice guidance.

- OpenShift Container Platform v4 introduces a compliance operator *[31]* to help ensure compliance and detect gaps with RHEL CoreOS based hosts in general. The **National Institute of Standards and Technology** (**NIST**) organization, among other things, also provides utilities and guidance for robust compliance practices that enterprises strictly adhere to as a general principle. The OpenShift Compliance Operator provides the right level of automation to help enterprises leverage the NIST OpenSCAP utility, for example, for scanning and enforcement of policies in such complex, dynamic environments.

Cluster admins are generally tasked with the responsibility of ensuring that the cluster is set up appropriately to meet compliance standards. It is also important that there is sufficient monitoring of privileged actions in the cluster, whether in the infrastructure or at higher levels, such as with Cloud Pak for Data. The next section identifies additional configuration that would be needed to ensure the reporting of all privileged activities.

Auditing

Auditing provides a way for enterprises to prove compliance and to monitor for irregularities.

Auditing can be enabled at various levels in the stack:

- At the operating system level *[32]*, RHEL provides mechanisms for the enterprise to monitor violations of any security policies. The audit system logs access watches for any sensitive or critical file access on that host.

- At the OpenShift Container Platform level *[33]*, Administrators can configure auditing levels to log all requests to the Kubernetes API server by all users or other services.

- At the Cloud Pak for Data level, audit logging supports generating, collecting, and forwarding **Cloud Auditing Data Federation** (**CADF**) *[34]* compliant audit records for core platform auditable events.

You can configure IBM Cloud Pak for Data audit logging to forward audit records to your **security information and event management** (**SIEM**) solutions, such as **Splunk**, **LogDNA**, or **QRadar**.

The audit logging implementation uses Fluentd output plugins *[35]* to forward and export audit records. After you install IBM Cloud Pak for Data, the Audit Logging Service is not configured out of the box. External SIEM Configuration *[36]* can be added to the `zen-audit-config` Kubernetes configmap to define where the Audit Logging Service forwards all collected audit records.

While audit logging generally only covers specific privileged actions, there are typically additional regulatory requirements with regards to tracking access of data by users and applications. In the following section, we will see how that too can be implemented.

Integration with IBM Security Guardium

Guardium provides the ability to monitor data access and to ensure data security. IBM Security Guardium's **Database Activity Monitoring (DAM)** capability offers an advanced centralized audit repository, reporting mechanisms, as well as tools to support forensics and compliance audits. These capabilities are complementary to the system-level auditing and privileged action audit recording described in previous sections as they can expand to data protection as well.

Cloud Pak for Data enables integration with Guardium in two ways:

- **Auditing sensitive data access**: Cloud Pak for Data users with the Manage Catalog permissions have the authority to identify sensitive data, such as **Personally identifiable information (PII)**, and configure it to be monitored by Guardium [**39**]. Since Cloud Pak for Data's Watson Knowledge Catalog service can also scan and tag sensitive data, it makes it easier for authorized users to enable those data assets to be monitored by Guardium.

- Cloud Pak for Data facilitates the provisioning of databases as well. It is critical for regulatory compliance that access to such data also be monitored. This is facilitated by a service called the Guardium *E-STAP* [*40*] service that is provisioned on top of Cloud Pak for Data and configured to intercept traffic to DB2 databases. All access to the DB2 database is then only conducted through the E-STAP interface. This service is configured to work with the Guardium appliance, which enables access logging and reporting, and access policy enforcement.

Guardium is an established and mature offering in the data protection and compliance segment. Many enterprises already leverage Guardium to automate compliance controls in a centralized fashion and the ability to integrate with Guardium enables Cloud Pak for Data customers to accelerate compliance readiness efforts with regard to their own data, complementing what is offered by the Cloud Pak for Data platform and operating environment.

In this section, we explored what techniques are available for the enterprise to get ready for regulatory compliance. In general, Cloud Pak for Data provides elegant mechanisms for RBAC and for the enterprise to be able to configure or change out-of-the-box roles to support any regulatory compliance, such as separation of duties.

Compliance requirements vary quite a bit depending on the industry the customer is in, as well as geographical requirements. In the preceding section, we also identified the need to appropriately configure all elements of the stack, including OpenShift Container Platform and its host nodes, as well as Cloud Pak for Data itself for privileged action auditing, and other integration options, such as Guardium to support data access monitoring.

Summary

In this chapter, we discussed how Cloud Pak for Data is developed and the IBM secure engineering practices that are employed. We also looked at how container images are delivered from trusted locations for customers to pull from or download.

Once Cloud Pak for Data is installed, care must be taken to ensure that it is operated in a secure manner. We looked at constructs with which RHEL and OpenShift provide a secure foundation for workloads and the security posture of Cloud Pak for Data on OpenShift Container Platform.

We also looked at how administrators are tasked with the responsibility of managing user and group access to ensure authorization is enforced in addition to configuring appropriate identity managers for authentication, as well as defining appropriate roles or modifying out-of-the-box user roles.

Security is critical to any software installation and Cloud Pak for Data is no exception. This chapter, in addition to describing the technology, also provided pointers to practical mechanisms to secure and monitor each part of the stack to ensure the safe and reliable operation of Cloud Pak for Data in your enterprise.

In the next chapter, we will look at storage, yet another foundational and critical construct for the reliable operation of Cloud Pak for Data.

References

1. **Cloud Pak for Data Security Whitepaper**: https://community.ibm.com/community/user/cloudpakfordata/viewdocument/security-white-paper-for-cloud-pak?CommunityKey=c0c16ff2-10ef-4b50-ae4c-57d769937235&tab=librarydocuments

2. **Redbook**: [Security in Development: The IBM Secure Engineering Framework: [https://www.redbooks.ibm.com/redpapers/pdfs/redp4641.pdf]

3. **IBM PSIRT**: [IBM Security and Vulnerability management]: https://www.ibm.com/trust/security-psirt

4. The **Open Web Application Security Project (OWASP)**: `https://owasp.org/www-project-top-ten/`

5. **Red Hat Enterprise Linux (RHEL)** base images: `https://access.redhat.com/documentation/en-us/red_hat_enterprise_linux/8/html-single/building_running_and_managing_containers/index/#working-with-container-images_building-running-and-managing-containers`

6. **IBM Container Software Library Access keys**: `https://myibm.ibm.com/products-services/containerlibrary`

7. **Operator Lifecycle Manager**: `https://docs.openshift.com/container-platform/4.6/operators/understanding/olm/olm-understanding-olm.html`

8. **Host and VM security**: `https://docs.openshift.com/container-platform/4.6/security/container_security/security-hosts-vms.html`

9. **Security Enhanced Linux (SELinux)**: `https://access.redhat.com/documentation/en-us/red_hat_enterprise_linux/8/html/using_selinux/getting-started-with-selinux_using-selinux#introduction-to-selinux_getting-started-with-selinux`

10. **OpenShift Security Context Constraints**: `https://docs.openshift.com/container-platform/4.6/authentication/managing-security-context-constraints.html`

11. **A Guide to OpenShift and UIDs**: `https://www.openshift.com/blog/a-guide-to-openshift-and-uids`

12. Kubernetes volume permission and ownership change policy: `https://kubernetes.io/docs/tasks/configure-pod-container/security-context/#configure-volume-permission-and-ownership-change-policy-for-pods`

13. Using a custom TLS certificate for the **Cloud Pak for Data** service: `https://www.ibm.com/support/producthub/icpdata/docs/content/SSQNUZ_latest/cpd/install/https-config-openshift.html`

14. **Encrypting etcd**: `https://docs.openshift.com/container-platform/4.6/security/encrypting-etcd.html`

15. **Portworx volume encryption**: `https://docs.portworx.com/portworx-install-with-kubernetes/storage-operations/create-pvcs/create-encrypted-pvcs/`

16. **Encryption with OpenShift Container Storage**: `https://access.redhat.com/documentation/en-us/red_hat_openshift_container_storage/4.6/html-single/planning_your_deployment/index#data-encryption-options_rhocs`

17. **IBM Cloud File Storage encryption**: `https://cloud.ibm.com/docs/FileStorage?topic=FileStorage-mng-data`

18. Implementing **Linux Unified Key Setup-on-disk-format (LUKS)**: `https://access.redhat.com/solutions/100463`

19. **Anti-virus software for RHEL**: `https://access.redhat.com/solutions/9203`

20. Configuring Single Sign-on with **SAML** for **Cloud Pak for Data**: `https://www.ibm.com/support/producthub/icpdata/docs/content/SSQNUZ_latest/cpd/install/saml-sso.html`

21. **IBM® Cloud Platform Common Services Identity and Access Management**: `https://www.ibm.com/support/knowledgecenter/en/SSHKN6/iam/3.x.x/admin.html`

22. Integrating **Cloud Pak for Data** with Cloud Platform Common Services: `https://www.ibm.com/support/producthub/icpdata/docs/content/SSQNUZ_latest/cpd/install/common-svcs.html`

23. Disabling the out-of-the-box **Cloud Pak for Data** Admin user: `https://www.ibm.com/support/producthub/icpdata/docs/content/SSQNUZ_latest/cpd/admin/remove-admin.html`

24. Configuring the **Cloud Pak for Data** idle session timeout: `https://www.ibm.com/support/producthub/icpdata/docs/content/SSQNUZ_latest/cpd/install/session-timeout.html`

25. Managing Roles in **Cloud Pak for Data**: `https://www.ibm.com/support/producthub/icpdata/docs/content/SSQNUZ_latest/cpd/admin/manage-roles.html`

26. Pre-defined Roles and Permissions in **Cloud Pak for Data**: `https://www.ibm.com/support/producthub/icpdata/docs/content/SSQNUZ_latest/cpd/admin/roles-permissions.html`

27. Managing **Cloud Pak for Data** user groups: `https://www.ibm.com/support/producthub/icpdata/docs/content/SSQNUZ_latest/cpd/admin/manage-groups.html`

28. Generating **Cloud Pak for Data** API keys: `https://www.ibm.com/support/producthub/icpdata/docs/content/SSQNUZ_latest/cpd/get-started/api-keys.html`

29. **Cloud Pak for Data** secrets API: `https://www.ibm.com/support/knowledgecenter/SSQNUZ_3.5.0/dev/vaults.html`

30. Configuring RHEL for compliance: `https://access.redhat.com/documentation/en-us/red_hat_enterprise_linux/8/html/security_hardening/scanning-the-system-for-configuration-compliance-and-vulnerabilities_security-hardening#configuration-compliance-tools-in-rhel_scanning-the-system-for-configuration-compliance-and-vulnerabilities`

31. OpenShift Compliance Operator: `https://docs.openshift.com/container-platform/4.6/security/compliance_operator/compliance-operator-understanding.html`

32. **RHEL Auditing**: `https://access.redhat.com/documentation/en-us/red_hat_enterprise_linux/8/html/security_hardening/auditing-the-system_security-hardening`

33. Configuring OpenShift Audit Polices: `https://docs.openshift.com/container-platform/4.6/security/audit-log-policy-config.html`

34. **Cloud Auditing Data Federation (CADF)**: `https://www.dmtf.org/standards/cadf`

35. **Fluentd output plugins**: `https://docs.fluentd.org/output`

36. Exporting IBM Cloud Pak for Data audit records to your security information and event management solution: `https://www.ibm.com/support/pages/node/6201850`

37. Creating Custom **Security Context Constraints (SCC)** for Services: `https://www.ibm.com/docs/en/cloud-paks/cp-data/4.0?topic=tasks-creating-custom-sccs`

38. Pre-defined Roles and Permissions in **Cloud Pak for Data**: `https://www.ibm.com/docs/en/cloud-paks/cp-data/4.0?topic=users-predefined-roles-permissions`

39. Data Access Monitoring with Guardium: `https://www.ibm.com/docs/en/cloud-paks/cp-data/4.0?topic=integrations-auditing-your-sensitive-data-guardium`

40. Guardium E-STAP for supporting compliance monitoring and data security for databases in Cloud Pak for Data: `https://www.ibm.com/docs/en/cloud-paks/cp-data/3.5.0?topic=catalog-guardium-external-s-tap`

11
Storage

Cloud Pak for Data leverages the Kubernetes concept of **persistent volumes**, which is supported in OpenShift, to enable its various services so that it can store data and metadata. Cloud Pak for Data services can also connect to remote databases, lakes, and object stores, as well as remotely exported filesystems as sources of data. We will look at the concepts and technologies that power both in-cluster PVs and off-cluster external storage. By the end of this chapter, you will have learned about what options are available in various public and private cloud infrastructures, as well as how to best optimize data storage for your use cases. You will also learn how storage should be operationalized, especially to support the continuous availability of your entire solution.

In this chapter, we're going to cover the following main topics:

- Persistent volumes
- Connectivity to external volumes
- Operational considerations

Understanding the concept of persistent volumes

We will start by providing a quick introduction to the foundational storage concepts in Kubernetes and how Cloud Pak for Data leverages storage. Typical containers are stateless, which means that any files they create or any changes they make will be ephemeral; upon restarting the container, these changes would not be restored. Also, in-container files would be created in a temporary location on the host, which can cause contention with other, similar consumers of that location. In **Kubernetes**, the concept of volumes [1] enables the persistence of directory files that are needed for stateful containers. Such volumes are mounted in a Kubernetes Pod and, depending on the access mode associated with that volume, the same directory contents can be shared with other Pods running concurrently within that Kubernetes namespace.

In this section, you will learn what it means to provision entities known as **persistent volumes**, which provide storage for Cloud Pak for Data workloads and can be configured for performance, reliability, and scale.

Multiple technologies and vendors enable storage in Kubernetes; we will explore those currently certified by Cloud Pak for Data.

Kubernetes storage introduction

Kubernetes provides a construct called **persistent volume claims** (**PVCs**)[2] as an abstraction for Pods to mount block and file **persistent volumes** (**PVs**) [2] without needing vendor-specific handling. Storage vendors, typically introduced as operators, provide the necessary services to define **storage classes** [3] and **dynamic provisioners** [4] to create, initialize, and mount these volume filesystems in different Pods. While the storage software and configuration are vendor-specific, the concept of PVCs makes consuming storage a standard procedure that's immune from the technology that's used to provide that storage.

Note that Cloud Pak for Data requires the use of **dynamic provisioners**; that is, the ability to provision storage on-demand. Static provisioners, which predefine storage volumes, are insufficient as Cloud Pak for Data services are deployed via a dynamic, *plug-n-play* mechanism. Scaling out such services could also, in many classes, automatically require us to create new volumes.

PVs and storage classes are Kubernetes resources at a cluster-wide scope. However, PVCs are OpenShift project (Kubernetes namespace) scoped entities, which means they can only be associated with stateful sets and Deployments within the same namespace.

Types of persistent volumes

From Cloud Pak for Data's perspective, PVs can be classified as follows:

- **Dedicated volumes**: These are block stores that are mounted by only one Pod at a time. Access mode: **Read-Write-Once (RWO)**.

- **Shared volumes**: These are shared filesystems that are mounted read-write by many Pods in the same namespace. Access mode: **Read-Write-Many (RWX)**.

The following diagram shows an example of the **user-home** persistent volume, which is mounted by many different Pods at the same time:

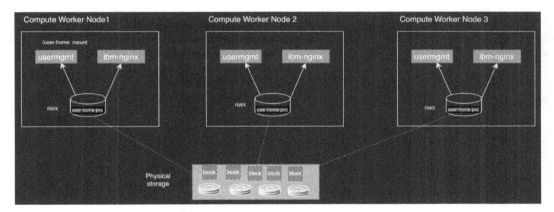

Figure 11.1 – The user-home shared volume

Depending on the storage technology being used, the blocks of data may be distributed across different disk devices and replicated for durability and resilience. Even if there are concurrent reads and writes from multiple Pods, these are typically synchronized across the different copies of the blocks. The storage drivers typically provide a POSIX [5] compliant filesystem interface and abstract all the physical aspects. Some vendors even provide caching techniques from a performance perspective.

Storage classes, represented by the `StorageClass` Kubernetes `Kind` resource, are used to reference each type of volume. When Cloud Pak for Data services are provisioned, they include PVC declarations, which are then processed by Kubernetes to provision and bind PVs.

For example, the `user-home` volume is requested by defining a `PersistentVolumeClaim` that resembles the following:

```
apiVersion: v1
kind: PersistentVolumeClaim
metadata:
  name: "user-home-pvc"
spec:
  storageClassName: {{ <user-selected-storage-class> }}
  accessModes:
    - ReadWriteMany
  resources:
    requests:
      storage: {{ default "100Gi" or <user-selected-size> }}
```

You will also notice that the PVC definition identifies parameters, such as the storage class to use and the storage size that's been requested. These are introduced via the installation options for the Cloud Pak for Data services.

The declarations in the *claim*, along with the class of storage and other attributes (such as `size` and `accessModes`), represent a *request* to Kubernetes to provision a volume for use by the Cloud Pak for Data service. Once the storage volume has been provisioned, Kubernetes then satisfies that claim by *binding* the provisioned PV to the PVC. This binding metadata then enables Kubernetes to always mount the same storage volume in the Pods, as well as for services to use the abstraction of the PVC to always refer to the volume, without needing to know the specifics of where the volume exists physically or the driver/storage technology behind it.

> **Note**
>
> You will not need to explicitly define the PVC yourself. The PVC will be declared during the Cloud Pak for Data service installation process.

`StorageClass` definitions can also identify the vendor-specific provisioner to be used. Some storage classes are predefined when the storage software is installed in the OpenShift cluster, while other vendors provide this for defining new storage classes, with custom parameters for fine-tuning volumes for specific use cases.

As an example, the `ocs-storagecluster-ceph-rbd` (RWO) and `ocs-storagecluster-cephfs` (RWX) storage classes are present as part of the OpenShift container storage installation, while Portworx defines storage classes explicitly with different sets of configurable attributes for performance or quality.

The Portworx Essentials for IBM installation package includes a list of storage classes that are used by Cloud Pak for Data services. Here is a snippet of one of the RWX classes; that is, `portworx-shared-gp3`:

```
apiVersion: storage.k8s.io/v1
kind: StorageClass
metadata:
  name: portworx-shared-gp3
parameters:
  priority_io: high
  repl: "3"
  sharedv4: "true"
  io_profile: db_remote
  disable_io_profile_protection: "1"
allowVolumeExpansion: true
provisioner: kubernetes.io/portworx-volume
reclaimPolicy: Retain
volumeBindingMode: Immediate
```

Here is a snippet of the `portworx-metastoredb-sc` RWO class:

```
apiVersion: storage.k8s.io/v1
kind: StorageClass
metadata:
  name: portworx-metastoredb-sc
parameters:
  priority_io: high
  io_profile: db_remote
  repl: "3"
```

```
    disable_io_profile_protection: "1"
  allowVolumeExpansion: true
  provisioner: kubernetes.io/portworx-volume
  reclaimPolicy: Retain
  volumeBindingMode: Immediate
```

Here, you can see that there are some vendor-specific parameters. Some are meant for performance optimization, while others are meant for redundancy (replication) and control, such as checking whether such provisioned volumes can be expanded.

Similarly, other storage provisioners, such as IBM Spectrum Scale, have similar configuration parameters as part of their storage classes.

Once Kubernetes provisions these PVs, as specified in the requested PVCs, Kubernetes uses the binding between the physical location of the storage and the PV construct to *abstract* how storage is accessed by individual Pods. These provisioned PVs are then mounted by Kubernetes upon startup of the individual Cloud Pak for Data Pods, using drivers provided by the storage vendors or the underlying technology.

As part of installing Cloud Pak for Data services, specific storage classes are selected by the user as installation parameters. The Cloud Pak for Data documentation [6] includes guidance on which types of storage classes are appropriate for each service [7].

Now, let's look a bit deeper into the storage topology itself, such as where the storage could be located and what it means for different Kubernetes Nodes to access this.

Broadly, we can classify the storage locations like so:

- **In-cluster storage**: This is where the physical storage devices/disks and the storage software control system are. They are used to provision and manage volumes in those devices, and they are also part of the same Kubernetes cluster as Cloud Pak for Data itself.

- **Off-cluster storage**: This is where the physical storage devices and the storage software is operated independently from the Kubernetes cluster where Cloud Pak for Data is expected to run. Only the storage drivers may exist on the Kubernetes cluster, as well as their provision and mount volumes from that remote storage system.

From a storage volume access perspective, various types of software drivers (software-defined storage) and plugin code exist in the Kubernetes ecosystem to support mounting volumes [12]. Some are called "in-tree" volume provisioners, which refer to code that natively exists in Kubernetes itself to work with such types of storage, while others are referred to as "out-of-tree" provisioners, which refer to code that exists outside of Kubernetes but can be plugged to work with different storage solutions. Recent versions of Kubernetes also include a driver standard called the **Container Storage Interface (CSI)** [13]. These provisioners present one or more storage classes that provide the storage abstraction needed for applications in Kubernetes. Hence, Kubernetes applications can be designed to work against the PVC and storage class standards and be completely independent of the actual physical location, the organization, or the drivers needed to work with storage.

In-cluster storage

This deployment option typically refers to co-locating storage and compute within the same Kubernetes cluster as Cloud Pak for Data services. Storage devices are attached to multiple compute worker or infrastructure Nodes, while a software-defined storage solution coordinates how the storage will be provisioned and consumed by mounting the volumes wherever they're needed. These Nodes are typically connected over a low latency and high throughput network inside the cluster. Such storage is only meant to be used by applications and workloads running inside that Kubernetes cluster.

Kubernetes natively includes the ability to mount volumes directly by using `hostPath` from the compute host [8], as well as node affinity mechanisms by using `local` volumes [9] to work around the logistical issues of scheduling Pods on the same workers as storage. However, these options do not provide dynamic provisioning support, and reliability is compromised when the underlying worker Nodes go offline. Also, OpenShift's `restricted` security context [11], by default, will not permit large numbers of such volumes from the host. The use of `hostPath` or `local` volumes also causes a tight coupling with Kubernetes applications in terms of storage. The portability of applications and the clean abstraction of storage classes and provisioning technologies is lost. For these reasons, Cloud Pak for Data does not rely on `hostPath` or `local` volume types.

Optimized hyperconverged storage and compute

This deployment model takes advantage of storage devices that are physically co-located in a *hyperconverged* fashion with compute. The storage is in the proximity of the same physical worker node as the compute, which enables optimization to occur, as well as workloads to distribute themselves to take advantage of such co-location [*10*]. This can be seen in the following diagram:

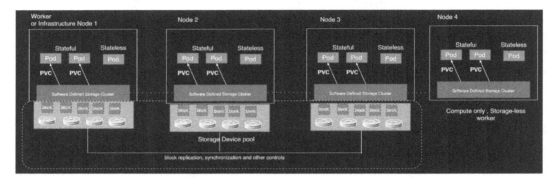

Figure 11.2 – Compute and storage in the same cluster

Note that the blocks of data may be distributed across devices on multiple Nodes, promoting durability and failure recovery for high availability. Some storage technologies can schedule the *co-location* of a Pod with physical storage and deliver optimization to provide a "local" mount, *without the overhead of network access* [*10*]. Others may still network mount volumes from the same host, but co-location still provides a performance advantage.

In the previous diagram, you will also notice that it is entirely possible to have Nodes that do not have storage but can mount volumes from other storage Nodes. This can potentially allow for partial co-location and independent scale-out of computing from storage.

Separated compute and storage Nodes

A slight variation of the in-cluster storage model is to have dedicated storage Nodes and dedicated compute Nodes. Storage Nodes can have disks but they only allow the storage software service Pods to run. That is, the Cloud Pak for Data Pods are run *separately* from storage, and each can be scaled independently:

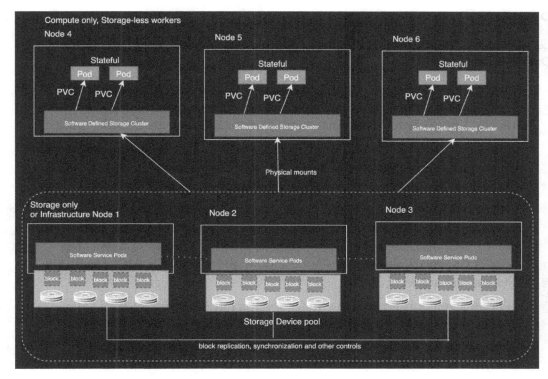

Figure 11.3 – Fully separated storage and compute Nodes

This organization can still be considered hyperconverged in the sense that the Kubernetes cluster supports both storage and compute. The private low latency network between the Nodes in the Kubernetes cluster would still be able to support access to storage from Pods in a performant manner.

This variation would not be able to support the possibility of a *"local"* mount optimization taking advantage of the physical co-location of storage blocks and compute, but it has the advantage of simplifying node operations and maintenance. This separation can help ensure the storage services are not impacted by a sudden surge (of noisy neighbor syndrome) from other Kubernetes workloads and vice versa.

Provisioning procedure summary

Let's take a look at the procedure for making such storage available to Cloud Pak for Data services:

1. First, you must install the storage software, including the associated provisioner, in the OpenShift cluster and configure the disk devices for locating the volumes. This is usually done through the storage vendor's operator and operator life cycle manager:

    Currently, Cloud Pak for Data supports **OpenShift Container Storage** (**OCS**) and Portworx Essentials for IBM/Portworx Enterprise. Spectrum Scale CSI Drivers and IBM Cloud File Storage are certified for a subset of its services.

    NFS is also supported, but NFS is not an in-cluster storage solution per se, but rather a protocol that allows us to access storage in a standard manner from remote file servers. NFS is used to provision volumes on other cloud storage providers such as **Google Cloud Filestore**, Azure locally redundant **SSDs**, **AWS Elastic File System**, and more.

    If needed, you can also label the Nodes or select infrastructure Nodes as storage Nodes, as opposed to co-locating storage with compute.

2. Next, define the storage classes as needed with references to the storage vendor-specific dynamic provisioner.

3. Install Cloud Pak for Data services and use the storage class names as parameters:

    PVCs and template declarations will include the storage class names for identifying the right type of volumes to provision.

    Individual Kubernetes Deployments and stateful sets specifications will associate these claims with mount points.

In this section, you learned how physical storage can compute and be part of the same Kubernetes cluster, as well as how to take advantage of such hyperconverged co-location. However, there is one distinct disadvantage – such a storage organization is considered scoped for containerized applications within that one cluster only. Two Kubernetes clusters would not be able to take advantage of a pool of storage devices, and the storage management operational expenses would double.

In the next section, we will look at a different organization, where storage is managed so that it supports multiple Kubernetes clusters, and perhaps even non-containerized storage requirements.

Off-cluster storage

In some situations, enterprises prefer to fully separate compute and storage, or even use a *centralized* storage management solution that is independently operated, to support the provisioning requirements of multiple clusters. This is quite similar to the *separated compute and storage (Figure 11.3)* model we described previously, except that the storage cluster is not part of the same OpenShift cluster. This also enables enterprises to leverage their existing investments in storage management solutions outside of Kubernetes.

It is important to note that the use of PVCs, with their abstraction of the location of physical storage, makes it possible for Kubernetes applications to seamlessly work in this case as well. Essentially, Cloud Pak for Data services mount these volumes without having to be know if the storage is external to the OpenShift cluster. Just selecting alternative storage classes during installation would make this possible.

In this deployment model, a sophisticated storage management solution exposes the ability to consume storage from many different Kubernetes clusters (and usually traditional VMs or bare-metal ones too), at the very least with the ubiquitous NFS protocol. Many storage solutions, such as IBM Spectrum Scale [15] and Red Hat OpenShift Container Storage [16], can also support this kind of topology using CSI drivers [13], without needing NFS.

NFS-based persistent volumes

Kubernetes supports the NFS protocol out of the box. However, it is meant to work with exported NFS volumes *outside* the cluster with vendor storage technology that can expose the NFS protocol. The Kubernetes project provides the NFS dynamic provisioner service, which can be used to identify an external NFS server and one exported filesystem, and then create PVs as sub-directories in that NFS exported filesystem on demand.

The following diagram shows an example of how a central storage management solution can expose volumes (either via the standard NFS protocol or other custom mechanisms):

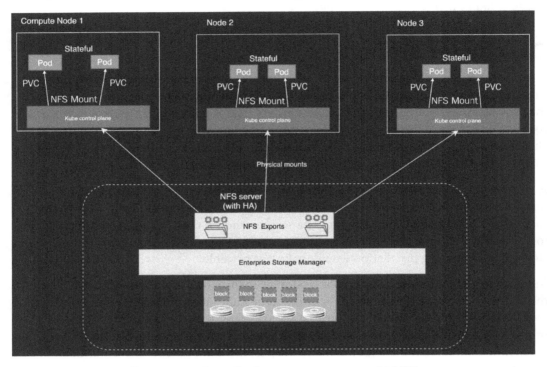

Figure 11.4 – Centralized storage management with NFS

Storage is expected to be provisioned dynamically as needed, by provisioners, to satisfy the PVCs defined in Kubernetes. Drivers (or, natively, NFS) support how these volumes are physically mounted in different Pods that have been scheduled on different Kubernetes compute workers.

For detailed guidance on how to use this provisioner and the storage class to define, see [14] in the *Further reading* section.

Note that the expectation, from a production perspective, is that this assigned NFS server does not become a single point of failure, that it is highly available, and that it supports automatic failover (that is, it can't be something as trivial as the NFS daemon on a single Linux system). If the NFS server fails, all the PVC mounts would become unusable. Hence, an enterprise-class resilient storage management solution is critical for ensuring overall reliability.

Operational considerations

In the previous sections, we discussed what it means to provide or consume storage either in-cluster or off-cluster for the containerized services in Cloud Pak for Data. However, it is also critical to make sure that the storage infrastructure itself is resilient and that operational practices exist to ensure that there is no data loss. Storage failover must also be taken into account from a **disaster recovery** (**DR**) standpoint.

In this section, we will take a quick look at some typical practices and tools we can use to ensure that the Cloud Pak for Data services operate with resilience, and that data protection is assured.

Continuous availability with in-cluster storage

A core tenet in Kubernetes is to ensure the continuous availability of applications. This includes load balancing and replicas for compute that can span multiple zones. Such zones could be *failure (or availability) zones* in public clouds or perhaps independently powered *racks* of hardware for an on-premises deployment [17].

In terms of in-cluster storage, this requires that the storage is also continuously replicated – this heavily depends on the storage vendor or technology in use. Here is an example topology for continuous availability, sometimes also referred to as **active-active** DR:

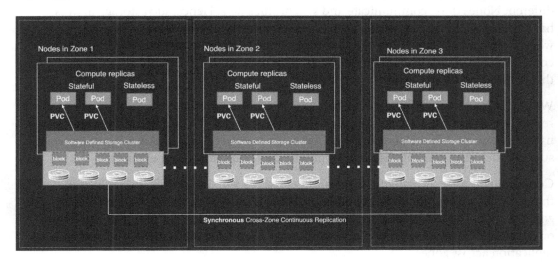

Figure 11.5 – Active-active DR

This model is sometimes referred to as a *stretched* cluster due to the assumption is that there is just one cluster, even though some Nodes may be located in different physical "zones."

The key here is to ensure that the storage blocks are replicated across zones in a near-real-time fashion. With this kind of deployment, stateful Pods would always have access to their PVs, regardless of which zone they are physically located in. The failure of one storage node/replica should not cause the entire application to fail. This is because other storage replicas should be able to serve the necessary blocks of data. The storage drivers in Kubernetes that support this notion of continuous availability completely abstract (via PVCs) where the data is physically picked up from, and they can fail over automatically to use storage from a different replica.

From a load balancing perspective, we typically consider that all compute Pods are always active and available for use, irrespective of the zone they are located in. Even if an entire zone becomes unavailable, only the compute and storage replicas become inaccessible, which means other replicas can take up the slack.

Such storage drivers are typically also configured to be aware of the cluster topology [18], for example, by the use of Kubernetes node labels to designate failure zones. This enables them to place PV replicas in different zones and thus offer resiliency in the case of the loss of one zone. A common problem with such a replication model is the "split brain" effect, where storage replicas have inconsistent data blocks because of failure situations where different Nodes don't communicate and synchronize data. Hence, such storage drivers have the sophistication to cordon off storage under these conditions, as well as identify when a replica has been offline for too long and needs to be updated to "catch up" or copy over data blocks when new storage replicas are added. This is critical to ensure that application data is always kept consistent in all its replicas.

While having a low latency network between such zones is critical, this multi-zone replication technique also requires such sophistication from the storage technology being used. For example, Red Hat OpenShift Container Storage (Ceph) and Portworx Enterprise support such requirements, as do most public cloud IaaS storage providers.

Continuous storage availability is technically possible with off-cluster storage as well, though it takes a bit more effort to align the storage zones with the Kubernetes cluster zones. External storage managers such as IBM Spectrum Scale, Dell EMC Isilon, NetApp onTap, and NAS-based solutions have traditionally provided mechanisms for synchronous replication across zones.

Data protection – snapshots, backups, and active-passive disaster recovery

It is critical for enterprises to ensure that consistent data and metadata backups occur at periodic intervals. Even if continuous availability can be achieved, having mechanisms for performing backups is essential. For example, having point-in-time backups help support "rollbacks" after data corruption incidents.

In some situations, offsite passive DR may be required. In this model, there isn't a stretch cluster or active Nodes, but a standby cluster that can be activated when there is a failure. If the DR site is geographically distant and network latency is not within the norms of a "stretch" cluster, or for a reliable performance of continuous replication for storage, then passive DR might be the only option available. In the case of a failover from the DR site, you would take previously backed up data and metadata from the active cluster and restore from backup it in the standby cluster. Then, you would designate the DR site cluster as the "active" cluster. The disadvantage here is the need to keep additional compute and storage resources in reserve, and that frequent backups are present, as well as mature operational playbooks to activate the cluster from those backups so that they meet the desired **Recovery Point Objective** (**RPO**) and minimize the downtime (**Recovery Time Objective**) the enterprise can tolerate.

Cloud Pak for Data, as a platform, provides options within the `cpd-cli` utility for data protection in general [19]. The `cpd-cli backup-and-restore` command enables two kinds of mechanisms to be leveraged:

- **Via a storage snapshot**: A snapshot of a PVC represents the state of that volume at that time. Snapshots are instantaneous because storage drivers do not take a full copy of the data blocks in that volume. The biggest advantage is that both taking a snapshot and restoring to that snapshot can be done very quickly.

    A typical installation of Cloud Pak for Data services uses many PVCs. This is why a "group" snapshot that takes a snapshot of multiple PVCs in that Kubernetes namespace is recommended. Note that CSI snapshots currently do not support group snapshots and that some vendors natively snapshot mechanisms. With Portworx [20], for example, a group PVC snapshot is done by temporarily freezing I/O on all those PVCs before triggering snapshots. The `cpd-cli` utility provides a `backup-restore snapshot` command that simplifies how Portworx snapshots are triggered and managed for all Cloud Pak for Data PVs.

Snapshots are commonly stored in the same devices, which means they are only as reliable as the devices themselves. Hence, storage vendors [20] also provide mechanisms for creating such snapshots for cloud storage, such as S3 object stores.

- **Via a copy of the available volume data**: This approach relies on rsync or a copy of some data from PVCs being sent to another PVC (typically, a remote storage volume) or an object store. The `cpd-cli backup-and-restore` utility has a Kubernetes service that allows us to copy data from PVs to the backup volume.

However, this technique requires all the services in that Kubernetes namespace to be put into "maintenance mode" (or fully quiesced) to prevent any I/O from occurring while the volume data is being copied. The advantage though is that this works for any kind of storage that's used, including NFS, for the PVs and does not require anything specific from the storage system.

Quiescing Cloud Pak for Data services

For application-level consistency, writes must be paused while backups are in progress. To support such a requirement, the `cpd-cli` utility includes a `backup-restore quiesce` command that tells services to temporarily put themselves into maintenance mode. This utility also scales down some of the Deployments to prevent any I/O activity. Once the backup is complete, the `backup-restore unquiesce` command is used to scale up services from maintenance. Note that while such a **quiesce** implies that Cloud Pak for Data is unavailable for that duration - that is, under maintenance mode – being able to achieve application-consistent backups is also important.

The **quiesce** mechanism can also be leveraged for external storage backups or snapshots, such as when NFS is used for PVs. For example, when IBM Spectrum Scale provides NFS volumes, you could quiesce the services in the Cloud Pak for Data namespace, then trigger a GPFS snapshot and unquiesce the services.

Db2 database backups and HADR

Ensuring that databases operate in a reliable and continuous manner is often critical to an enterprise. While write suspends/quiesce is supported, Db2 provides additional techniques that allow Db2 databases to be available online [22] for use, even when backup operations are in progress. This is also independent of the storage technology chosen and it can support incremental backups.

The Db2 **High Availability Disaster Recovery** (**HADR**) configuration [23] airs two Db2 database instances that could exist in either the same cluster (say, a different Kubernetes namespace) or even a remote cluster. In this configuration, data is replicated from the active source database to a standby target. There is an automatic failover to the standby when a site-wide failure occurs or when the active Db2 database is deemed unreachable. The Db2 HADR **Automatic Client Reroute** (**ACR**) feature is a mechanism where clients connecting to the failed Db2 database seamlessly get switched over to the new active Db2 database.

Kubernetes cluster backup and restore

The snapshot mechanisms and the `cpd-cli backup-and-restore` utility described previously can only copy data from PVCs and restore it in new PVCs. When restoring such a backup in a new cluster, the expectation is that the target cluster is set up in the same way the original cluster was. This means that the necessary Cloud Pak for Data services must be installed in the target cluster, and that all the Kubernetes resources are present before the data restoration process takes place.

The *Velero* project [24] introduced a mechanism for backing up all Kubernetes resource artifacts as well. This allows the target cluster to be reconstructed with ease before data restoration occurs. The OpenShift API for Data Protection Operators [25], which was introduced by Red Hat for OpenShift clusters, combined with *Ceph-CSI* drivers, provides a way to better operationalize cluster-level backup and restore. It also supports additional use cases, such as to duplicate or migrate clusters.

Summary

In this chapter, we introduced the concept of PVs in Kubernetes and how Cloud Pak for Data works with different storage technologies and vendors. We also discussed storage topologies and vendors or technologies that may exist in different enterprises.

This chapter also stressed the importance of data protection and the operationalization of backup and restore, as well as DR practices. Ensuring that you have a deep understanding of storage ensures that Cloud Pak for Data services and the enterprise solutions that are built on top of them operate with both resiliency and the desired level of performance.

Further reading

1. *Kubernetes Volumes*: https://kubernetes.io/docs/concepts/storage/volumes/

2. *Kubernetes Persistent Volumes and Claims*: https://kubernetes.io/docs/concepts/storage/persistent-volumes/

3. *Storage Classes*: https://kubernetes.io/docs/concepts/storage/storage-classes/

4. *Dynamic volume provisioning*: https://kubernetes.io/docs/concepts/storage/dynamic-provisioning/

5. *Portable Operating System Interface (POSIX)*: https://pubs.opengroup.org/onlinepubs/9699919799/nframe.html

6. *Cloud Pak for Data Documentation – Storage Considerations*: https://www.ibm.com/docs/en/cloud-paks/cp-data/3.5.0?topic=planning-storage-considerations

7. *Cloud Pak for Data documentation – Hardware and Storage Requirements per Service*: https://www.ibm.com/docs/en/cloud-paks/cp-data/3.5.0?topic=requirements-system-services#services_prereqs__hw-reqs

8. *Kubernetes hostPath-Based Volumes*: https://kubernetes.io/docs/concepts/storage/volumes/#hostpath

9. *Kubernetes Local Volumes*: https://kubernetes.io/docs/concepts/storage/volumes/#local

10. *Portworx Hyperconvergence Using the Stork Scheduler*: https://docs.portworx.com/portworx-install-with-kubernetes/storage-operations/hyperconvergence/

11. *OpenShift Security Context Constraints for Controlling Volumes*: https://docs.openshift.com/container-platform/4.7/authentication/managing-security-context-constraints.html#authorization-controlling-volumes_configuring-internal-oauth

12. *Container Storage Interface Volume Plugins in Kubernetes Design Doc*: https://github.com/kubernetes/community/blob/master/contributors/design-proposals/storage/container-storage-interface.md

13. *Container Storage Interface Specification*: https://github.com/container-storage-interface/spec/blob/master/spec.md

14. *NFS External Provisioner*: https://github.com/kubernetes-sigs/nfs-subdir-external-provisioner

15. *IBM Spectrum Scale Container Storage Interface*: `https://www.ibm.com/docs/en/spectrum-scale-csi`

16. *Deploying OpenShift Container Storage in External Mode*: `https://access.redhat.com/documentation/en-us/red_hat_openshift_container_storage/4.6/html/deploying_openshift_container_storage_in_external_mode/index`

17. *Kubernetes – Running in Multiple Zones*: `https://kubernetes.io/docs/setup/best-practices/multiple-zones/`

18. *Portworx – Cluster Topology Awareness*: `https://docs.portworx.com/portworx-install-with-kubernetes/operate-and-maintain-on-kubernetes/cluster-topology/`

19. *Backup and Restore*: `https://www.ibm.com/docs/en/cloud-paks/cp-data/3.5.0?topic=recovery-backing-up-restoring-your-project`

20. *Portworx PVC Snapshots*: `https://docs.portworx.com/portworx-install-with-kubernetes/storage-operations/kubernetes-storage-101/snapshots/#snapshots`

21. *OpenShift Container Storage Snapshots*: `https://access.redhat.com/documentation/en-us/red_hat_openshift_container_storage/4.6/html/deploying_and_managing_openshift_container_storage_using_google_cloud/volume-snapshots_gcp`

22. *Db2 Backup*: `https://www.ibm.com/docs/en/cloud-paks/cp-data/3.5.0?topic=up-online-backup`

23. *Db2 HADR*: `https://www.ibm.com/docs/en/cloud-paks/cp-data/3.5.0?topic=db2-high-availability-disaster-recovery-hadr`

24. *Velero for Kubernetes Backup and Restore*: `https://velero.io/docs/v1.6/`

25. *OpenShift API for Data Protection*: `https://github.com/konveyor/oadp-operator`

12
Multi-Tenancy

Multi-tenancy is a software architecture design where multiple users or organizations share the same instance of the software and its underlying resources. This is a standard practice with **Software-as-a-Service (SaaS)** offerings, such as Cloud Pak for Data as a service in the IBM Cloud, but the interpretation of tenancy and how this tenancy is implemented can vary widely on-premises or in private clouds. **Security** and **compliance** requirements in the enterprise, as well as the *quality of service assurances*, play a significant role in deciding how much can be shared.

Nowadays, it is a fundamental expectation that any true cloud-native platform stack enables at least some level of multi-tenant sharing out of the box. Supporting a new project or initiative can be accelerated just by reusing the existing shared infrastructure. Trivially, you could assume assigning independent **virtual machines (VMs)** to each tenant would be sufficient to achieve multi-tenancy – after all, the underlying bare metal resources are indeed being shared. But it is not cost-effective to provide dedicated clusters of VMs to every department in an enterprise; the operational maturity needed to manage multiple clusters or installations of software alone would become prohibitively expensive.

Most enterprises, whether they operate in a hosted public cloud infrastructure or on-premises, prefer some level of sharing between tenants with assurances of consistent performance, security, and **privacy**. It is no longer considered adequate just to assign VMs to tenants and, with the advent of containerization techniques, more granularity is now possible with multiple tenants even sharing the same VM. This also makes security problems more complex since even the operating system gets shared, with processes from different tenants running side by side.

IBM uses sophisticated multi-tenancy techniques in its cloud for the Cloud Pak for Data SaaS offering, and the architectural underpinnings of the Cloud Pak for Data software stack can also enable enterprises to support multiple tenants on the same OpenShift Kubernetes cluster. *Chapter 9, Technical Overview, Management, and Administration,* introduced how multi-tenancy can be achieved with the judicious use of **Kubernetes namespaces**, where a single OpenShift cluster is shared between multiple tenants. There needs to be careful deliberation with regards to how to implement multi-tenancy in a secure and reliable fashion while being mindful of your enterprise's requirements, policies, and the **service guarantees** expected by tenants.

In this chapter, we're going to be covering the following topics:

- Tenancy considerations
- Architecting for multi-tenancy
- A review of the tenancy requirements
- In-namespace sub-tenancy with looser isolation

Tenancy considerations

We will start by looking closely at what multi-tenancy implies, why tenancy is important, and, in general, what architectural choices, trade-offs, and compromises regarding expenses need to be made when planning for such shared deployments.

Designating tenants

Each enterprise usually has a very different interpretation when it comes to identifying their tenants, but by and large, a tenant is defined as a select group of users granted specific privileges to access a shared software installation. When tenants represent different companies altogether, users belonging to one tenant are generally unaware of the existence of other tenant users. However, even such isolation can be relaxed when it comes to users within the same enterprise, and employees in the same company may be members of multiple tenant "accounts."

Here are some example situations where a Cluster Operations team may want to designate "tenants" that share the same OpenShift cluster:

- **Development** versus **staging** versus **production environments**: For strict regulatory compliance reasons, different users are granted different access privileges to each environment. Production environments may be maintained at a different "stable" software version level compared to development, which may even be using beta-level versions. Resources may be more likely to be re-assigned in favor of production workloads when needed at the expense of the development or staging environments – that is, performance assurances, compute quotas, and throttling policies may differ for each "tenant."

- **Departmental tenants**: In the same enterprise, different organizations may have different compute needs and policies with regards to data or other resources. Chargeback requirements would also imply that the compute resources being utilized by the different departments be meterable independently and **service-level agreements (SLAs)** can be guaranteed.

- **Tenants managed by SaaS and solution vendors**: With Cloud Pak for Data, just as with Cloud Pak for Data SaaS, we see partners and other providers operate clusters for their own customers in turn, essentially as SaaS offerings. There are even solution providers who develop higher-level solutions using Cloud Pak for Data capabilities. Security is paramount since these service providers expect to run workloads from completely *different companies* in the same shared cluster. Apart from the need to support individual per-tenant access management, chargeback, and SLAs, the requirement of isolating such tenants is generally much more stringent. Even one tenant being aware of the presence of another tenant or one tenant's microservice being able to perform a network lookup or access another tenant's microservice would not be tolerated.

Now that we have an idea of who or what can be tagged as tenants, in the next section, we will let's take a look at what it means for the enterprise and IT operations to support safely sharing clusters with different tenants.

Organizational and operational implications

To support multi-tenancy, there needs to be an understanding of the structural aspects within the enterprise, starting with clearly identifying the roles and responsibilities of various teams and the need to define SLA style contracts to set expectations. SLAs become important as a way to provide assurances of performance and reliability when many units in that enterprise (or even external parties) claim a share of compute resources, storage, and software services.

In this section, we will explore aspects of hosting and managing shared clusters in general and supporting tenant users in particular.

Administering shared clusters

One tenant user cannot be assigned as a superuser of the shared OpenShift cluster when there are other tenants present. Not only would it represent a conflict of interest, but access to such a tenant's assets or compute resources is expected to be private. This is also relevant regarding Cloud Pak for Data on OpenShift, and appropriate controls exist to assure clear separations of duties.

Hence, there needs to be a special independent group – say, the **Cluster Operations** team – that is responsible for all IT needs, including scaling out clusters by adding new nodes or ensuring there's a fair share of compute resources or throttling, as per the tenant service agreements all maintenance requirements. It would be the responsibility of this team to manage and monitor the OpenShift cluster, ensure processes are implemented for handling alerts, and that the problems at hand are resolved. Most importantly, the operations team would be signing a contract of sorts with the tenants (a SLA) to offer assurances regarding performance, availability, and security on this shared cluster.

Note, however, that the "cloud-native" pattern implies that individual tenants (or authorized users) expect some sort of *self-service*, including the ability to install or upgrade software services and troubleshoot issues without the need to involve superuser cluster admins.

> **A note on the role of the operations team**
>
> With Cloud Pak for Data, this cluster *operations* team essentially manages OpenShift and Kubernetes, its underlying hosts, as well as key infrastructures such as storage and network constructs. Most importantly, they decide what it means to identify and govern tenancy in the OpenShift cluster.

Within each tenancy, specific users can be granted administrative privileges that have been *scoped* to their instance to give the tenant some measure of self-service. This allows tenant admins to perform some of the Cloud Pak for Data management functions, including managing user access themselves and troubleshooting, as described in *Chapter 9, Technical Overview, Management, and Administration.*

A note on the role of tenant users

Including this one Tenant Admin role, all users from a tenant group should be considered *end users*; that is, personas who are focused on *consuming* the capabilities offered by the Cloud Pak for Data services and not responsible for managing the installation or the cluster infrastructure by themselves.

Tenant Admins are also end users but have a slightly more important role to play in the interest of **self-service** management – they bridge the gap between the operations team and the consumers of the Cloud Pak for Data instance.

Onboarding tenants

In such multi-tenanted deployments, there needs to be a formal process of identifying tenants, granting them the appropriate (limited) access to the assigned resources, and metering their usage. In Cloud Pak for Data, the prescribed approach is for the operations team to create unique OpenShift project namespaces, with the appropriate **ResourceQuotas**, for provisioning Cloud Pak for Data services.

Depending on the policies with regards to Kubernetes access, the OpenShift cluster admin could grant a tenant admin user limited RBAC to that OpenShift project namespace. This would allow such tenants to provision Cloud Pak for Data services (or upgrade them) as needed.

Note

The use of **operators**, as described in *Chapter 9, Technical Overview, Management, and Administration,* is designed to help both the operations team (operators are, after all, operational knowledge coded into the software) and the tenants themselves. Cluster admins introduce any operators as needed for their tenant use cases (and associated container images) and keep them up to date. When tenants need to provision a service in their namespace, they would only need to "declare" their need for it using an appropriate custom resource specification, without needing to be experts in understanding how to automate the process of installing (or upgrading) any service.

Alternatively, *without* granting tenants Kubernetes access, the operations team could take up more responsibility and even provision and configure selected Cloud Pak for Data services as needed for the tenant's use cases. Once done, the operations team could just assign a tenant user *Administrator* access to that instance of Cloud Pak for Data. The operations team could even decide to assign Cloud Pak for Data privileges at a very granular level, withholding some privileges. For example, a tenant Cloud Pak for Data administrator may not be granted the *Manage Configurations* or *Manage Platform Health* permissions, perhaps only *View Platform Health*. The tenant admin is usually granted permissions to manage user access on their own, but they may be denied the ability to change the configured authentication mechanism. Such elevated privileges may continue to be retained by the operations team. Cloud Pak for Data's concept of roles and permissions (as described in *Chapter 10, Security and Compliance*) provides the flexibility needed for such tenant-level policies to be established.

Note that from a self-service perspective, granting *Manage Platform Health* permissions to the Cloud Pak for Data (tenant) admins is common and highly recommended as it helps the tenant manage their workloads and deployed services on their own, as well as handle a fair share of compute resources using quotas, as described in the *Resource management* section of *Chapter 9, Technical Overview, Management, and Administration*. These quotas would remain scoped only to that namespace and well within the upper bound defined by the **ResourceQuota** that had been set for that namespace by the operations team.

> **Note**
>
> In summary, onboarding tenants in the context of Cloud Pak for Data implies that the operations team's cluster admins, as well as the tenant admins, assume specific responsibilities.

The operations team has the responsibility of doing the following:

- Ensuring enough compute and storage is available for the tenant, perhaps expanding the OpenShift cluster if needed.

- Making operators available in the cluster and mirroring the appropriate images to a registry that can be accessed by the Kubernetes cluster.

- Creating at least one OpenShift Project (Kubernetes namespace) for that tenant.

- Defining the appropriate ResourceQuota on the tenant's namespace(s) based on service assurance requirements or SLA contracts with the tenants.

- Identifying a tenant admin or "owner" and granting them OpenShift Kubernetes access and limited namespace administration RBAC to the newly created namespaces, but only if the enterprise policy permits this.

- Installing the Cloud Pak for Data control plane and the appropriate services via custom resources (if direct Kubernetes access is not permitted for tenants), configure authentication, and designate one of the tenant users as a Cloud Pak for Data admin with the appropriate permissions.

- Providing support for day 2 operations, such as upgrading (updating operators), scaling out, and establishing backup procedures.

The tenant administrator has the responsibility of doing the following:

- Installing the Cloud Pak for Data control plane via custom resources, if direct Kubernetes access and namespace RBAC has been granted.

- Ensuring security by granting sign-in access to authorized users, defining the appropriate roles for users and user groups, as well as nominating alternate tenant "admins."

- Publishing the necessary URL for accessing Cloud Pak for Data to users who have been granted access.

- Defining user groups when needed for different users to collaborate in Cloud Pak for Data.

- Installing other services via custom resources, as needed for specific use cases, if Kubernetes access is permitted or working with the operations team to make those services available.

- Monitoring and managing the Cloud Pak for Data platform within that tenant namespace scope, including scaling out/in services and establishing usage quotas based on the expected workload.

- Backing up (and restoring when needed) if Kubernetes access is granted; otherwise, plan a backup strategy with the operations team.

In this section, we provided some insights into how the need to support multiple tenants with assurances of security and reliability could define how the enterprise IT and business teams become organized. There are, of course, variations to such arrangements, especially when there are different legal entities or companies involved and the tenancy evolves over time, with new requirements and new kinds of tenants making an appearance. Hence, the approach to and even planning for multi-tenancy must be flexible, aspects of which we will explore in the next section.

Architecting for multi-tenancy

In this section, we will explore what it means to implement multi-tenancy with Cloud Pak for Data, starting with how to plan out the topology on an OpenShift cluster and how to separate the tenants within the same cluster.

Achieving tenancy with namespace scoping

As we established earlier, assigning a unique OpenShift project (a Kubernetes namespace) to each unique tenant is the recommended approach for Cloud Pak for Data multi-tenant deployments.

The following diagram shows how multiple tenants can be sandboxed into individual OpenShift projects and the logical separation between the **Operations zone** (which would be managed by the cluster administrators) and the **Tenants zone** (somewhere the tenant users are potentially given access to):

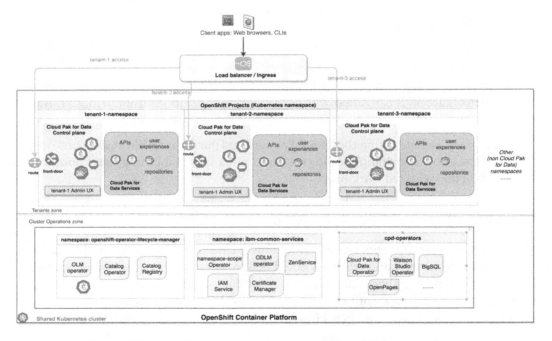

Figure 12.1 – Individual tenant namespaces in a shared Kubernetes cluster

The **Cluster Operations zone** includes all the management functions provided by OpenShift and Kubernetes (not everything is shown in the preceding diagram for brevity). The **Operator Lifecycle Manager** (**OLM**) provides operators and services for installing and managing Cloud Pak for Data itself. It is also very much relevant for achieving multi-tenancy as it provides the right abstraction to extend the cluster and enable different capabilities (via service operators) to be made available in the cluster. To meet separation of duties requirements, OLM can be fully controlled by the cluster admins and integrated into their CI/CD processes, with tenants, including their admins, denied access.

The **Cloud Pak Foundational Services** (**CPFS**) are commonly located in the *ibm-common-services* namespace. This includes the *namespace-scope* operator, which is instrumental in ensuring that the Cloud Pak for Data Operators are constrained to only managing specific tenant namespaces and not *all* namespaces. There isn't a need for these operators to watch over *non*-Cloud Pak for Data namespaces anyway and for security reasons (principle of least privilege), it is best not to even grant these operators that privilege. While the Cloud Pak for Data service operators can be co-located with CPFS in *ibm-common-services*, they can also be installed in a different *cpd-operators* namespace, as shown in the preceding diagram. For example, you may choose to use *cpd-operators* when there are other Cloud Paks present in the same cluster. While it is necessary for the CPFS operators to manage all Cloud Pak namespaces, it is not necessary for Cloud Pak Data operators and by isolating them to *cpd-operators*, you can ensure they are scoped to only the Cloud Pak for Data namespaces.

Note

As depicted in the preceding diagram, each tenant namespace has its own private copy of the Cloud Pak for Data control plane scaled out as needed, including a front door with its own private ingress/route. OpenShift enables the exposure of this front door for access via unique tenant URLs. This enables client apps to target the appropriate tenant instance to connect to from outside the cluster.

Ensuring separation of duties with Kubernetes RBAC and separation of duties with operators

Let's explore what roles operators play in this namespace-based multi-tenancy approach:

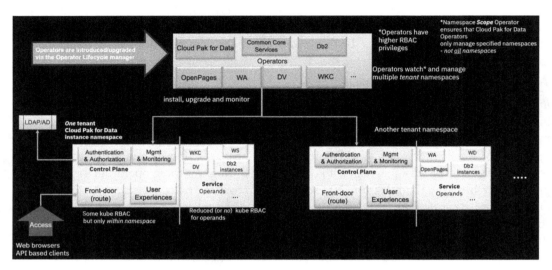

Figure 12.2 – Operators managing multiple namespaces

Custom resources are declarations that trigger operators – deployments that reside in their own namespace – to install the control plane and any Cloud Pak for Data services that are needed by a tenant. Operators, via calls to the Kubernetes API, create the Kubernetes artifacts (operands) and upgrade them, based on the custom resource specifications. Introducing a new service or capability in Cloud Pak for Data for a tenant would be as simple as declaring an appropriate Custom Resource.

A note on the importance of operators

A key advantage of this operator-based pattern is that the Kubernetes privileges (in the form of RBAC) that are assigned to these tenant service accounts can be reduced, or even completely removed, since the bulk of the Kubernetes API interactions are managed by the operators. These service accounts are scoped to just that tenant's namespace and not at the cluster level anyway. Even then, the overall security posture in these shared multi-tenant clusters can be improved just by reducing the tenant service account's RBAC privileges.

Future versions of Cloud Pak for Data services will see a further reduction in the need for RBAC as more and more functions move over to the operator side.

Securing access to a tenant instance

The following diagram provides a closer look at what objects exist inside a tenant's namespace:

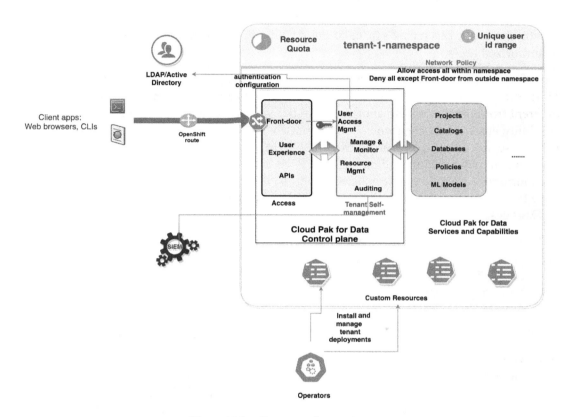

Figure 12.3 – One tenant's namespace

Recall from *Chapter 9, Technical Overview, Management, and Administration*, where we discussed the role of the Cloud Pak for Data control plane and its location within a Kubernetes namespace to facilitate a single access point (the "front door") for that tenant's users, thereby enabling an integrated user experience, including self-service management capabilities, for any designated tenant admin. The preceding diagram further emphasizes what this implies from a single tenant's perspective.

Authentication and auditing for each tenant can be configured independently. The signing key that's used for generating bearer tokens is unique to each tenant. Hence, using an authentication token from one tenant's instance would not work if you wish to gain access to another. Authorization, in the form of *roles and permissions*, can help tenant admins grant authorization to that tenant's users in a granular fashion. Objects and capabilities inside this tenant's installation of Cloud Pak for Data would only be available to authorized users.

OpenShift also automatically generates a unique *UID* range for each project. Hence, the Linux processes within one Cloud Pak for Data tenant namespace run with a UID that is different from processes from another namespace. This, coupled with **RHEL's SELinux** capability, ensures that even when pods from two tenants are scheduled in the same Linux host, these processes are prevented from interacting with one another. Thus, a malicious process that may have infected one tenant's Pod, or user errors that may have accidentally been introduced, would not be able to impact the Pod processes from another tenant. Note that some Cloud Pak for Data v4 services rely on a fixed UID via the use of a custom SCC (as described in *Chapter 10, Security and Compliance*).

With network policies, which permit access within the namespace, microservices can communicate with one another. However, access from another tenant's microservices can be effectively blocked, except for those requests that come in through the front door, and only those with the right authorization.

Choosing dedicated versus shared compute nodes

In OpenShift, SELinux and the unique tenant UID for Linux processes provide a significant amount of security when we have processes from different tenants executing side by side, sharing the same operating system kernel runtime. With namespace ResourceQuotas and the ability for Cloud Pak for Data to monitor and enforce usage quotas, there are additional controls to ensure that a noisy tenant workload does not adversely affect another tenant and that SLAs can be assured in the cluster.

However, there may be security policies in the enterprise that do not permit co-locating one tenant's processes with another or SLAs that demand some sort of physical separation between tenants. Obviously, since assigning independent Kubernetes clusters to individual tenants is operationally cost-prohibitive, the OpenShift container platform provides an option to support such requirements.

A cluster administrator could introduce a *NodeSelector* annotation to OpenShift projects to designate which compute worker hosts (identified by labels) are to be used for scheduling the pods for that tenant. This ensures there are dedicated nodes for that tenant in an otherwise shared OpenShift cluster. As part of a tenant's onboarding process, a cluster admin could decide whether to use shared compute hosts or assign dedicated hosts, just by using this annotation. While with some services, such as databases, it is possible to use the Kubernetes concept of Pod-level **Node affinity**, it is typically much easier to assign all nodes for a tenant to dedicated nodes if necessary.

Reviewing the tenancy requirements

In *Chapter 9, Technical Overview, Management, and Administration*, we briefly introduced what key attributes need to be addressed to support multi-tenancy with Cloud Pak for Data.

In a way, we can relate these requirements to tenancy for applications in traditional operating system environments, albeit at a larger scale. A Unix or Linux system would be set up with all the software needed, and the superuser ("root") has the responsibility of managing this. The root user would authorize users to access that system and assign them to different user groups ("tenants"). The use of trivial filesystem permissions (group IDs and user IDs) enables these users to work together in their tenant group and be somewhat isolated from other tenants. Users could execute and access applications that they (or their group) have been granted access to, and operating system security primitives (such as SELinux and AppArmor) ensure that cross-process access is strictly controlled.

> **Note**
>
> The Kubernetes cluster is merely the *new* "operating environment" and supports many of the same paradigms as the operating systems, just at a higher level. Cluster admins are the "root" users, and they can grant various role-based access rights to other users. From a multi-tenancy perspective, such admins now have additional operating environment primitives to leverage, such as OpenShift security context constraints, isolation of tenants in virtual projects/Kubernetes namespaces, and granular RBAC, as described in *Chapter 10, Security and Compliance*.

In this section, we will *assess* the following key tenancy requirements:

- Isolation
- Security and compliance
- Self-service management

We will look at them in terms of deploying Cloud Pak for Data instances in different Kubernetes namespaces.

Isolating tenants

It's a basic requirement that users from one tenant account are completely unaware of another tenant's existence. Even when a user is part of two tenant accounts – say, an employee who is assigned access to two departmental instances – there would be an explicit mechanism to "switch" from one tenant context to another – perhaps even different login procedures. In a situation where Kubernetes microservices associated with different tenant instances run in the same cluster, in close network proximity, and with data potentially being accessible, it becomes important to establish some boundaries and enforce strict separation.

Let's look at what Cloud Pak for Data and OpenShift offer in terms of achieving isolation.

Configuring authentication

Cloud Pak for Data provides the ability for each tenant installation in a namespace to decide how to authenticate the users in that tenant group. For example, one tenant instance could be configured to authenticate against one LDAP server, while another instance could be configured to use another. Alternatively, different LDAP (sub)groups could be used for different instances.

It might be desirable to configure **single sign-on** (**SSO**) against the *same enterprise* SSO provider for all (departmental) tenant instances. However, even in this case, access to individual Cloud Pak for Data instances can be controlled so that only specific users with roles in that instance have access.

Authorizing user access

In *Chapter 10*, *Security and Compliance*, we looked at *roles and permissions* in detail, including how they may be customized to suit the enterprise. Roles are a cornerstone when it comes to managing user access, even at very granular levels. Based on policies or practices, the cluster admin may even choose to modify the out-of-the-box Cloud Pak for Data "administrator" role, perhaps to remove the ability to (re)configure authentication or manage health, leaving only the ability to view monitoring information.

Apart from the platform (control plane), every service or capability requires authorization. For example, different groups of users could be designated as editors in a project or a catalog, while others, even if they have been successfully authenticated, could be denied access to that resource. In the same way, but with more granularity, individual users or user groups could be granted *select* access to a particular virtual table or view as part of the data virtualization service, while others may be denied that, just the same as with traditional relational databases.

Assuring isolation of compute resources

Each tenant in their own namespace gets *unique copies* of the microservices they need and can independently *scale* them out to suit their expected workloads or concurrency requirements. ResourceQuotas at the namespace level ensure that compute resources can be *reserved* for each tenant. Enterprises can also choose to assign dedicated hosts for a tenant to further guarantee isolation and the availability of compute resources.

As we discussed previously, OpenShift project UIDs and SELinux can assure that processes – even those colocated on the *same* host – are isolated.

Externalizing routes and entry points

Each tenant has a unique front door and route. With OpenShift and the use of external load balancers, it is even possible to issue completely new URLs with host and domain names that are appropriate for that tenant. Each front door route can also be associated with a tenant-specific TLS certificate, advertising which specific tenant "owns" that instance of Cloud Pak for Data.

This results in an easy and elegant way to isolate client traffic to the individual tenant Cloud Pak for Data instances.

Ensuring privacy with storage

The use of Kubernetes namespaces also guarantees that persistent volumes are only available within that namespace scope. As described in *Chapter 11*, *Storage*, **persistent volume claims (PVCs)** are only valid within one namespace and are used to mount storage volumes onto the Cloud Pak for Data microservice pods. Even if a cluster-wide shared storage solution is in use, using PVCs guarantees that one tenant's volumes cannot be accessed by another tenant's microservices.

There are additional guardrails that some storage solutions offer, including the use of per-tenant encryption keys to further improve security with shared storage solutions. Even with remote NFS storage servers configured to be shared with multiple tenants, the adoption of PVCs ensures that every volume is fully separated into sub-directories in that remote NFS volume. PVCs guarantee that only the appropriate sub-directories from that remote volume can be mounted into a tenant's pods. The OpenShift project's guarantee of unique UIDs implies that even on that NFS volume filesystem, permissions can be maintained to isolate ownership (and access rights) to individual tenant namespaces.

Tenant security and compliance

Beyond the ability to isolate tenants, enterprises need to guarantee security and privacy are maintained on that shared cluster. Regulatory compliance requirements need to be met, not just by the cluster operations team but also individual tenants themselves. In *Chapter 10*, *Security and Compliance*, we explored some of these aspects already, including techniques for supporting compliance verifications.

In this section, we will summarize how those techniques continue to apply to individual tenants in their quest to ensure compliance. We will also highlight the best practices and security features that are available in Cloud Pak for Data in this regard.

Securing access in a shared OpenShift cluster with RBAC

With only operators needing to retain the majority of Kubernetes RBAC, and with the ability to separate tenants into private namespaces, we already have a strong guarantee that any RBAC that's been given to individual tenant service accounts or tenant users is fully scoped to their specific Cloud Pak for Data instance. This elegantly ensures that one tenant can be denied access to another tenant's namespace.

It is also common to establish *separation of duties* for regulatory compliance, even *within* the namespace, to define distinct roles that do not overlap. For instance, you could define an "Access Management" role that is only permitted to manage access to the platform while ensuring it has no access to data or catalogs, or even the ability to set data access policies.

It is also important to note that only *owners* or *admins* of the appropriate objects should have the ability to give access to others. For example, a database administrator or data virtualization administrator decides who can connect to that instance of the service in the first place. The role of a Cloud Pak for Data platform administrator does *not* automatically imply that the user is a superuser and has access to everything, especially data. With such access authorization controls, it is also possible to even prevent cluster administrators from the operations team from gaining access to data.

Securing network access

In the previous sections, we established how the front door and unique routes help guarantee isolation for accessing individual tenant Cloud Pak for Data instances from *outside* the cluster. Kubernetes includes the concept of network policies, a concept very similar to firewall rules, that further strengthen the security of a tenant's instance from intrusive access from *within* the cluster.

Network policies can be authored to permit access between microservices within the same namespace and to deny, by default, access from other namespaces. Additional *allow* policies are defined, as an exception to the *deny all* rule, to permit HTTPS access to each tenant's front door and between the operators' namespace and each tenant's namespace. The *deny* policy blocks traffic between tenants, except for their front doors. Interactions between tenants, *if at all* desired, would need to be done via the authorization checks in their respective front doors, much like an external client accessing from outside the cluster.

Hence, while some access still needs to be permitted between the Operator and tenant namespaces, by carefully crafting network policies, tenants can ensure they are secure from intrusions either from outside or within the Kubernetes cluster.

Auditing

There always needs to be checks and balances with regards to how software services are operated in such multi-tenant architectures. From a regulatory compliance perspective, auditing is one powerful tool that enterprises use to provide assurances of safe and monitored operations. In *Chapter 10, Security and Compliance*, we discussed the many mechanisms that the cluster operations team can employ to ensure security and compliance.

Tenants typically need to adhere to compliance or monitor the usage of their instance of Cloud Pak for Data from a security perspective. Such tenants should not have access to auditing or monitoring records across the whole cluster, just those scoped to their instance.

Cloud Pak for Data includes an instance scoped auditing service, as described in *Chapter 10, Security and Compliance*. Each tenant admin or security focal, if properly authorized, can configure just their instance of Cloud Pak for Data to forward all relevant audit events to **security information and event management** (**SIEM**) systems. Even if the same SIEM is used by all tenants and by the cluster operations team, the ability to identify each tenant uniquely can be maintained, and tenant-specific auditing reports can be published.

Self-service and management

It is not cost-effective for the Cluster Operations team to hold every tenant user's hand for day-to-day operations. From a compliance perspective, it is not usually permitted for such cluster admins to be tightly embedded into a tenant's business. Besides, with cloud-based systems, there is every expectation that tenants can self-serve themselves for most requirements, without needing IT support. However, overall cluster security is the responsibility of the Cluster Operations team, and they need to be able to have assurances themselves that the tenants cannot breach beyond their instance scope.

Hence, it becomes a fine balance between maintaining rigid security constraints and the flexibility needed for tenants to manage their instances themselves every day. In this section, we will look at the capabilities that OpenShift and Cloud Pak for Data offer to help enterprises balance this need.

Provisioning and upgrades – enabling a choice of services and versions

Different tenants need different sets of Cloud Pak for Data services to be deployed. They may even need to "pin" them to specific versions from a stability perspective – typically, production environments are maintained at well-tested levels, while there is more flexibility for development environments.

With the judicious use of operator custom resources, individual tenants can decide to install and upgrade each service by themselves. With services that support the concept of *service instances*, such as Db2, appropriately authorized tenant users can even spin up a database instance on their own, without the need for Kubernetes access.

Hence, cluster admins could grant tenant admins privileges in their namespace to empower them to provision Cloud Pak for Data services on their own. In turn, tenant admins can authorize a subset of tenant users to be able to provision service instances on their own within the confines of that Cloud Pak for Data instance.

Monitoring

With tenant instance scoped monitoring and management, cluster admins can delegate day-to-day monitoring to individual tenant admins. Cluster admins still retain the rights to control the namespace as a whole, especially by establishing resource quotas, and tenant admins operate within those boundaries to decide how to monitor and throttle usage inside their tenancy.

The Cloud Pak for Data Monitoring and Alerting frameworks can help tenant admins understand utilization within their instance, as well as decide how much compute they must allocate to individual use cases or services.

Operating and maintaining tenant resources

Cluster admins can also empower tenant admins to perform any needed day 2 maintenance on their own. Patches or version upgrades of individual services (just like installation and provisioning, as described previously) can be triggered via custom resources by the tenant admins themselves, if that authority has been granted. custom resources can also be easily leveraged to scale out services in that specific instance of Cloud Pak for Data (`scaleConfig` fields in v4 work similar to how the `cpd-cli scale` command works with v3.5, except that it's driven by an Operator) with just namespace scope privileges being needed.

The monitoring and management experience provides an easy way for tenant admins to manage their own Cloud Pak for Data instances, including identifying and correcting failing services. They have access to individual logs and the ability to grab advanced diagnostics for IBM support, all without needing their cluster admins to engage.

Backup (and restore, when needed) is also a key concern for tenants. While cluster admins ensure the reliability of storage and other infrastructure across all tenants, backing up tenant-specific data and metadata can also be entrusted to tenant admins.

The *cpd-cli backup-restore* utility, as described in *Chapter 11, Storage*, requires scoped Kubernetes RBAC, and this utility can be leveraged by the tenant admins. Note, however, that with storage snapshots and other advanced techniques to move storage to remote sites, some initial setup may still be needed that the Cluster Operations team must manage. Importantly, each tenant's backup images can still be uniquely identifiable, isolated from other tenants' backups, and kept protected.

A summary of the assessment

In the previous sections, we discussed how Cloud Pak for Data and OpenShift can support the key requirements of multi-tenancy.

The following table provides a quick summary of how each requirement can be met and the mechanisms available that can be leveraged to implement that requirement:

Tenancy Aspects	Mechanism
Isolation	*Individual Kube Namespace quotas/OpenShift Projects for each Cloud Pak for Data instance*
1. Users – Authentication, Roles and Authorization 2. Compute Resources 3. Access to Instance 4. Storage Usage	✓ If needed, can support *different* LDAP/AD or SAML configurations *per* instance (each instance has its own unique user management service) ✓ Namespace quotas – set by the cluster admin. ✓ Unique DNS–based routes for each instance. ✓ Completely separated PVCs (in different namespaces) with different PVs – no sharing *across* instances.
Security	*RBAC privileges & Service Accounts scoped only to that namespace*
1. Restricting Kube Cluster-Level Privileges 2. Network Access 3. Auditing	✓ Service accounts and role binding – scoped to Project namespaces. ✓ Kube network policies for hardening *like a firewall around an RHOS Project namespace.* ✓ *Per-tenant* SIEM forwarding supported.
Management	*Administration privileges scoped to a namespace*
1. Monitoring/Metering 2. Ops – Backup/DR, Scale, Patch, Upgrade 3. Serviceability 4. Service Installations	✓ Different tenant "Admin users for different instances. Tenants are managed independently." ✓ Ops privileges scoped to that OpenShift project's namespace. ✓ Serviceability utilities only capture diagnostics that are appropriate for that instance. ✓ Different instances can have different sets of services at different versions scaled independently from other tenants.

Figure 12.4 – Summary of the mechanisms and techniques that support multi-tenancy

As a result of this approach to implementing multi-tenancy, enterprises and cluster administrators can decide how best to mix and match different tenants on the same shared OpenShift-based infrastructure.

The following diagram shows the topology of different tenants using different versions of Cloud Pak for Data services:

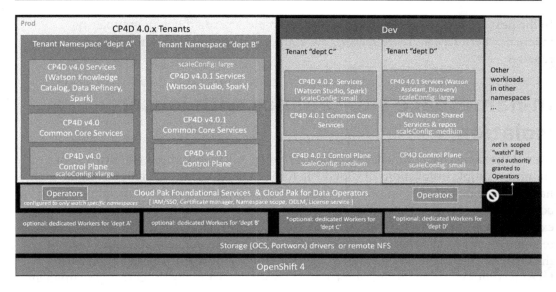

Figure 12.5 – An example of tenant diversity

The Cluster Operations team could choose to use the same OpenShift cluster for both production and development tenants, as shown in the preceding diagram. Special tenants could be assigned dedicated compute nodes (with increased cost) if appropriate. Each tenant could have deployed and pinned specific versions of the control plane itself and different Cloud Pak for Data services, as well as having configured them at different scale sizes.

> **A note on operand version choices for tenants**
>
> While operands themselves could be pinned to different versions, it is highly recommended that the set of Cloud Pak for Data operators be continuously upgraded to the latest release level, even in *air-gapped* data centers, to ensure the latest critical defect fixes are applied and any security vulnerabilities are addressed. Cloud Pak for Data operators support pinning operands to specific versions, even if the operators themselves get upgraded.
>
> Here, these operators support provisioning and maintaining *multiple versions* of the operands, and they can support both the tenant's choice to *automatically* upgrade operands to the latest released version or to *retain* (pin) operands to a specified version, with admins deciding when to manually upgrade.

It is important to note that not all enterprises may have the same requirements or policies. Some may require even more stringent isolations, say to the point of needing dedicated compute nodes for each tenant. Production tenants may need to be located on completely different OpenShift clusters, even in different networks or data centers. Other tenants may have relaxed requirements – for example, tenants that are designated as *development* environments typically do not have audits enabled.

In the next section, we will look at one such example, where a different granularity in tenancy permits a lot more sharing, but at the expense of security and isolation requirements.

In-namespace sub-tenancy with looser isolation

Even if the same OpenShift cluster is shared, there is still the overhead and expense of running multiple installations of Cloud Pak for Data services for each tenant. While namespace-based separation provides much more assurance in terms of meeting security, compliance, and performance SLA guarantees, it comes with additional expense. For example, we would need to install separate copies of the Kubernetes deployments in each tenant namespace, and that implies more oversight and maintenance personnel (even if the Operator pattern makes it easier to upgrade, scale, and so on). Separate copies also take up additional compute and storage resources, thus increasing the cost of procurement in the first place, as well as ongoing operational expenses.

In some cases, in the interests of reducing expenditure, some enterprises may choose to share more resources between *tenants*, even if it means losing flexibility or reduced isolation. In this section, we will explore what it means to share the same Cloud Pak for Data instance among multiple *sub-tenants*.

Approach

This borrows from the idea of **sublets**, where a primary tenant decides to share resources with sub-tenants. In this example, the primary tenant is assigned a Cloud Pak for Data instance by the Cluster Operations team, as usual. The tenant admin then decides to define their own sub-tenants to grant access to.

The primary tenant admin typically does the following:

- Creates a Cloud Pak for Data *user group* for each sub-tenant. Users and LDAP groups can be mapped in to uniquely identify each sub-tenant user.

- Assigns the appropriate platform roles – perhaps only giving the authority to create projects and catalogs.

- Provisions instances of services such as databases and grants access rights to individual sub-tenant groups – and even delegates ownership/administration to individual tenant users.

Hence, this approach primarily relies on access authorization and user groups to isolate tenant users from each other, as much as possible, intending to reduce compute resource cost. This implies that some operational compromises must be made, especially in terms of security, as well as reduced guarantees with regards to performance and perhaps even reliability.

Assessing the limitations of this approach

It is important to note that this notion of sub-tenancy would not work for all enterprises. For example, only one authentication mechanism can be configured and the expectation is that all users, irrespective of which sub-tenant user group they belong to, must be authenticated by that mechanism. Users would be able to "see" each other across different sub-tenant groups.

Authorizations, especially at a service instance level or for individual projects or catalogs can, to a great extent, provide secured sharing for the same Cloud Pak for Data instances among sub-tenants. However, it is to be noted that all the necessary isolations and security needs may not be met. It is also imperative that the privileges of a tenant admin be carefully managed (or fully retained by the Cluster Operations team for themselves so that they don't need to designate a separate tenant admin). Typically, deeper scrutiny of the roles that will be assigned to each tenant user group is required.

The following table provides an idea of how this sub-tenancy approach rates in terms of the key tenancy requirements:

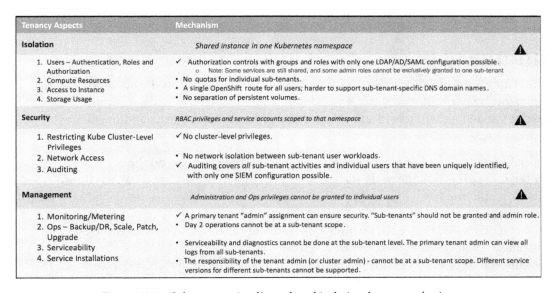

Figure 12.6 – Sub-tenancy implies reduced isolation, but more sharing

Some Cloud Pak for Data services are currently singletons, which means that it would not be possible to provision multiple instances in the same namespace and assign them to different sub-tenant user groups. For example, unlike Db2, data virtualization, transformation, and the enterprise catalog are singleton and meant to be shared services. This has other implications, such as a data virtualization admin or engineer being able to grant access to all data being virtualized, irrespective of which sub-tenant owned that data. This might mean an organizational change too, where there are designated virtualization admins or engineers who perform the needed operations *on behalf of the sub-tenants*, since it would not be secure to grant such roles to the sub-tenants themselves.

This approach also places more responsibilities on the primary tenant admin (and/or cluster admins). In the absence of being able to enforce a per sub-tenant quota, the tenant admin must be diligent in monitoring and initiating scale out and conflict resolutions preemptively, such as terminating runaway workloads from one sub-tenant that could impact others. Such tenant admins, for example, would need to weigh the impact on other sub-tenants when upgrading a service, as requested by just one sub-tenant.

A note on evaluating the suitability of the sub-tenant approach

As outlined in the preceding table, there are a few limitations to consider, especially in terms of ensuring isolation and separation of concerns between sub-tenants or even confidently offering the assurance of performance SLAs to each sub-tenant. The assessment for adopting this sub-tenancy needs to happen carefully at each Cloud Pak for Data service by service case or per capability, and the trade-off in terms of cost-savings versus isolation and reliability must be well understood, with the right organizational risk mitigations in place.

Summary

This chapter continued covering the foundational concepts we introduced in *Chapter 9, Technical Overview, Management, and Administration*, in terms of what it means to operate Cloud Pak for Data in a multi-tenanted fashion, with security and reliability in mind.

We elaborated on the key tenancy requirements, and then focused on the namespace per tenant approach as the best practice for supporting multiple Cloud Pak for Data tenants within the same shared OpenShift Kubernetes cluster. We also described how this recommended approach, along with important organizational structures, helps achieve the key tenancy requirements, starting with how Kubernetes namespaces and OpenShift projects, as a concept, help with enforcing tenant isolations. This chapter also reinforced the importance of the Operator pattern to help in improving the security posture and the different capabilities in the control plane that support tenancy, along with assurances of security and meeting performance SLAs.

As we reach the end, let's reflect on what we have explored in these chapters. We started by providing an introduction to the "AI ladder," a prescriptive approach to accelerating value from data and infusing trusted AI into your business practices. We emphasized the need for a sophisticated Data and AI platform, and how Cloud Pak for Data brings that platform to any cloud, as well as on-premises.

You learned how the platform enables key capabilities that help users collaborate to modernize data access, organize data for governance and insight, and use advanced analytics, which leads to elegantly infusing AI into enterprise processes. We also dove deep into the technical aspects of what powers the platform by taking a quick look under the hood. We presented architectural designs that emphasized that Cloud Pak for Data is a scalable and resilient Data and AI platform, implemented on top of a mature Kubernetes technology stack and built to ensure safe, elastic, cost-efficient, and reliable multi-tenanted operations.

In conclusion, Cloud Pak for Data is a feature-rich platform, built on modern cloud-native architectural principles, that aims to provide the right foundation for enterprises to elegantly deliver impactful data and AI solutions in the hybrid cloud.

Packt.com

Subscribe to our online digital library for full access to over 7,000 books and videos, as well as industry leading tools to help you plan your personal development and advance your career. For more information, please visit our website.

Why subscribe?

- Spend less time learning and more time coding with practical eBooks and Videos from over 4,000 industry professionals

- Improve your learning with Skill Plans built especially for you

- Get a free eBook or video every month

- Fully searchable for easy access to vital information

- Copy and paste, print, and bookmark content

Did you know that Packt offers eBook versions of every book published, with PDF and ePub files available? You can upgrade to the eBook version at packt.com and as a print book customer, you are entitled to a discount on the eBook copy. Get in touch with us at customercare@packtpub.com for more details.

At www.packt.com, you can also read a collection of free technical articles, sign up for a range of free newsletters, and receive exclusive discounts and offers on Packt books and eBooks.

Other Books You May Enjoy

If you enjoyed this book, you may be interested in these other books by Packt:

Learn Amazon SageMaker

Julien Simon

ISBN: 9781800208919

- Create and automate end-to-end machine learning workflows on Amazon Web Services (AWS)
- Become well-versed with data annotation and preparation techniques
- Use AutoML features to build and train machine learning models with AutoPilot
- Create models using built-in algorithms and frameworks and your own code
- Train computer vision and NLP models using real-world examples
- Cover training techniques for scaling, model optimization, model debugging, and cost optimization
- Automate deployment tasks in a variety of configurations using SDK and several automation tools

Mastering Azure Machine Learning

Christoph Körner, Kaijisse Waaijer

ISBN: 9781789807554

- Setup your Azure Machine Learning workspace for data experimentation and visualization
- Perform ETL, data preparation, and feature extraction using Azure best practices
- Implement advanced feature extraction using NLP and word embeddings
- Train gradient boosted tree-ensembles, recommendation engines and deep neural networks on Azure Machine Learning
- Use hyperparameter tuning and Azure Automated Machine Learning to optimize your ML models
- Employ distributed ML on GPU clusters using Horovod in Azure Machine Learning
- Deploy, operate and manage your ML models at scale
- Automated your end-to-end ML process as CI/CD pipelines for MLOps

Packt is searching for authors like you

If you're interested in becoming an author for Packt, please visit authors.packtpub.com and apply today. We have worked with thousands of developers and tech professionals, just like you, to help them share their insight with the global tech community. You can make a general application, apply for a specific hot topic that we are recruiting an author for, or submit your own idea.

Share Your Thoughts

Now you've finished *IBM Cloud Pak for Data*, we'd love to hear your thoughts! Scan the QR code below to go straight to the Amazon review page for this book and share your feedback or leave a review on the site that you purchased it from.

https://packt.link/r/1-800-56212-8

Your review is important to us and the tech community and will help us make sure we're delivering excellent quality content.

Index

A

www.ingramcontent.com/pod-product-compliance
Lightning Source LLC
Chambersburg PA
CBHW062100050326
40690CB00016B/3158